Ernst Schering Research Foundation Workshop 25
Novel Approaches to Treatment of Osteoporosis

Springer-Verlag Berlin Heidelberg GmbH

Ernst Schering Research Foundation
Workshop 25

Novel Approaches
to Treatment
of Osteoporosis

R.G.G. Russell, T.M Skerry, U. Kollenkirchen
Editors

With 84 Figures, Some in Color and 10 Tables

 Springer

Series Editors: G. Stock and U.-F. Habenicht

ISSN 0947-6075
ISBN 978-3-662-09009-1

CIP data applied for

Die Deutsche Bibliothek – CIP-Einheitsaufnahme
Schering-Forschungsgesellschaft <Berlin>: Ernst Schering Research Foundation Workshop.

ISSN 0947-6075
25. Novel approaches to treatment of osteoporosis. - 1998
Novel approaches to treatment of osteoporosis: with 10 tables / R.G.G. Russell...ed.

(Ernst Schering Research Foundation Workshop; 25)
ISBN 978-3-662-09009-1   ISBN 978-3-662-09007-7 (eBook)
DOI 10.1007/978-3-662-09007-7

Typesetting: Data conversion by Springer-Verlag

SPIN: 10657883   13/3135–5 4 3 2 1 0 – Printed on acid-free paper

Preface

Osteoporosis occurs as a result of loss of bone mass and deterioration of structure that leads to an increased risk of fractures. The consequences of this inappropriate bone loss are a growing drain on the resources of developed countries and are likely to become still more pervasive as the population ages and increasing areas of the third world adopt western occupations, diets and lifestyles. The challenges that are generated by osteoporosis are numerous but fall into three broad categories: diagnosis, prevention and treatment of pathological bone loss. Diagnostic methods currently allow the identification of individuals who have lost significant amounts of bone, which puts them at risk for fracture, but at present there is no good predictive method which allows intervention to begin very early in the loss phase of osteoporosis. This limits the utility of antiresorptive strategies and drugs. After diagnosis of osteoporosis by bone mineral measurement or the occurrence of a fracture, the current generations of single or combined therapies are highly effective in preventing further bone loss, but cause only a small and insufficient increase of bone mass.

In the future, one attractive solution to this problem would be to develop rapid predictive diagnostic tests which could be used to screen the population so that antiresorptive therapies could be started before significant bone was lost. However apart from the considerable problems of developing an accurate cheap and acceptable test with a significantly greater predictive power than any currently available, such a solution will not solve today's problems of the numerous people who have lost and will lose bone before they are diagnosed. There is a need therefore to develop treatments which increase bone mass, either in combination with an antiresorptive regimen or alone. Currently there are only a few agents which have the ability to stimulate new bone formation, despite the intensive research efforts to identify treatments of that sort.

In this two-day workshop held in November 1997 in Berlin, Germany, we set out to promote an interdisciplinary discussion between experts from different fields of research related to osteoporosis. The aim of this meeting was to present, discuss and assess novel approaches in the treatment (not prevention) of this disease. This included discussions of techniques, pathogenic mechanisms, and bone-active agents. Although the workshop could only focus on some important novel approaches in osteoporosis research and had to disregard others, the authors managed to cover a large amount of exciting research in this field.

We are indebted to the Ernst Schering Research Foundation for providing us with the necessary resources and help to organize this workshop. We wish to thank the authors for their contributions and the lively discussions at the workshop. We also thank the chairpersons, Dr. Reinhold Erben, University of München, Dr. Franz Theuring,Humboldt University of Berlin, and Drs. Heiner Fritzemeier, Martin Haberey and Rudolf Knauthe from Schering AG of Berlin, for guiding the sessions at the workshop.

Berlin, September 1998   *Uwe Kollenkirchen*
                                                 R. Graham G. Russell
                                                 Timothy M. Skerry

Table of Contents

List of Editors and Contributors

Editors

R.G.G. Russell
Human Metabolism and Clinical Biochemistry,
Sheffield University Medical School, Beech Hill Road,
Sheffield S 10 2RX, UK

T.M. Skerry
Department of Biology, University of York, PO Box 373,
York YO1 5YW, UK

U. Kollenkirchen
Schering AG, FC/HT Research, Müllerstr. 178, 13342 Berlin, Germany

Contributors

M. Amling
Department of Trauma Surgery, Hamburg University, Lottestrasse 59,
22529 Hamburg, Germany

J.N. Beresford
School of Pharmacy and Pharmacology, University of Bath, Claverton Down,
Bath BA2 7AY, UK

B. F. Boyce
Department of Pathology, University of Texas Health Science Center,
Floyd Curl Drive, San Antonio, Texas 78284-7750, USA

A. Dai
Department of Pathology, University of Texas Health Science Center,
San Antonio, Texas 78284-7750, USA

G. Delling
Department of Bone Pathology, Hamburg University, Martinistrasse 52,
20246 Hamburg, Germany

K. Dobson
Bone Formation Group, Department of Human Metabolism
and Clinical Biochemistry, University of Sheffield Medical School,
Beech Hill Road, Sheffield S10 2RX, UK

R.L. Duncan
Department of Orthopaedic Surgery, Biomechanics and Biomaterials
Research Center, Indiana University, Purdue University at Indianapolis,
Indianapolis, IN 46202, USA

E. F. Eriksen
Aarhus Bone and Mineral Research Group, Department of Endocrinology
and Metabolism, Aarhus University Hospital, Aarhus Amtssygehus, Denmark

A.E. Grigoriadis
Departments of Orthodontics and Pediatric Dentistry
and Craniofacial Development, UMDS Guy's Hospital,
Guy's Tower Floor 28, London Bridge, London SE1 9RT, UK

M.W. Hentz
Department of Bone Pathology, Hamburg University, Lottestrasse 59,
22529 Hamburg, Germany

D.E. Hughes
Department of Pathology, University of Sheffield, Beech Hill Road,
Sheffield S10 2RX, England

C. Jefferiss
School of Pharmacy and Pharmacology, University of Bath, Claverton Down,
Bath BA2 7AY, UK

S. Jones
Bone Formation Group, Department of Human Metabolism and
Clinical Biochemistry, University of Sheffield Medical School,
Beech Hill Road, Sheffield S10 2RX, UK

B.L. Langdahl
Aarhus Bone and Mineral Research Group, Department of Endocrinology
and Metabolism, Aarhus University Hospital, Aarhus Amtssygehus, Denmark

L.E. Lanyon
The Royal Veterinary College, University of London, Royal College Street,
London NW1 0TU, UK

P. Liu
Bone Formation Group, Department of Human Metabolism
and Clinical Biochemistry, University of Sheffield Medical School,
Beech Hill Road, Sheffield S10 2RX, UK

P.J. Marie
INSERM Unit 349, Cell and Molecular Biology of Bone and Cartilage,
2, rue Ambroise Paré, Lariboisière Hospital, 75475 Paris Cedex 10, France

D. Miao
Bone Formation Group, Department of Human Metabolism
and Clinical Biochemistry, University of Sheffield Medical School,
Beech Hill Road, Sheffield S10 2RX, UK

F.M. Pavalko
Department of Physiology, Biomechanics and Biomaterials Research Center,
Indiana University, Purdue University at Indianapolis, Indianapolis,
IN 46202, USA

M. Priemel
Department of Bone Pathology, Hamburg University, Lottestrasse 59,
22529 Hamburg, Germany

L. Reading
Bone Formation Group, Department of Human Metabolism
and Clinical Biochemistry, University of Sheffield Medical School,
Beech Hill Road, Sheffield S10 2RX, UK

J. Screen
School of Pharmacy and Pharmacology, University of Bath, Claverton Down,
Bath BA2 7AY, UK

A. Scutt
Bone Formation Group, Department of Human Metabolism
and Clinical Biochemistry, University of Sheffield Medical School,
Beech Hill Road, Sheffield S10 2RX, UK

C. Shui
Bone Formation Group, Department of Human Metabolism
and Clinical Biochemistry, University of Sheffield Medical School,
Beech Hill Road, Sheffield S10 2RX, UK

T.M. Skerry
Department of Biology, University of York, PO Box 373, York YO1 5YW
UK

K. Stewart
School of Pharmacy and Pharmacology, University of Bath, Claverton Down,
Bath BA2 7AY, UK

K. Still
Bone Formation Group, Department of Human Metabolism
and Clinical Biochemistry, University of Sheffield Medical School,
Beech Hill Road, Sheffield S10 2RX, UK

A. Sunters
Craniofacial Development Department, Guys Hospital, Guys Tower 28,
London Bridge, London SE1 9RT, England

L.J. Suva
Department of Bone and Cartilage Biology,
Smith Kline Beecham Pharmaceuticals, 709 Swedeland Road,
King of Prussia PA 19406, USA

C.H. Turner
Department of Orthopaedic Surgery, Biomechanics
and Biomaterials Research Center, Indiana University,
Purdue University at Indianapolis, Indianapolis, IN 46202, USA

S. Walsh
School of Pharmacy and Pharmacology, University of Bath, Claverton Down, Bath BA2 7AY, UK

K. R. Wright
Division of Human Anatomy, Loma Linda University, Loma Linda, California 92350, USA

L. Xing
Department of Pathology, University of Texas Health Science Center, San Antonio, Texas 78284-7750, USA

1 The Advantages and Limitations of Cell Culture as a Model of Bone Formation

K. Dobson, S. Jones, P. Liu, D. Miao, L. Reading, C. Shui, K. Still, and A. Scutt

1.1 Introduction

Despite its rather unglamorous image, bone is a remarkably complex and versatile organ, the correct functioning of which is essential for our general health. Bone formation comprises a complex but ordered sequence of events, beginning with the proliferation of chondrogenic and osteogenic precursor cells followed by their subsequent differentiation, ultimately leading to extracellular matrix (ECM) maturation and mineralization. Under normal circumstances, this process is tightly controlled by a combination of the endocrine system and locally acting growth factors. Bone can also respond to physical stimuli and removal of this stimulus, as experienced by astronauts in zero gravity or by patients undergoing extended bed-rest, results in a net loss of bone. Sometimes, the mechanisms controlling bone growth are defective, e.g., in nonunion of fractures and osteoporosis. Sometimes bone development is defective, as in achondroplasia, caused by a lack of chondrocytes in the growth plates of long bones resulting in short limbs, or gigantism, which is due to the failure of the growth plates to fuse at puberty allowing the bones to grow throughout life.

It is therefore of obvious importance to understand how bone functions both in health and disease. Unfortunately, bone does not lend itself easily to investigation. This is because bone is a complex network consisting of various types of cells existing in a defined ECM. The cells interact both with each other and with the ECM and when cells are removed from this network they cease to function normally. Previously, bone cell differentiation has been mainly studied using histological methodology in either whole embryos or organ cultures. Whilst this has provided much information regarding the temperospatial relationships of the various cells, the complexity of the systems excludes the possibility of elucidating the detailed molecular mechanisms involved in bone development and calcification. Similarly, whilst cell culture techniques have given us much information regarding the control of cell proliferation, the results cannot easily be applied to bone as a whole organ. Recently, a number of in vitro models have been established which recreate discrete elements of this network. Although osteoblasts (OBs) at various stages of differentiation are present, these models are not as complex as whole organs and, because of that, defined aspects of bone formation can be investigated at the cellular and molecular level.

1.2 Bone Formation and Regeneration In Vivo

Before we can evaluate the effectiveness of in vitro systems, we must first consider bone formation in vivo. Bone is a complex organ of osseous, cartilaginous, fibrous and hematopoietic tissue. During the development of new bone, these four tissues must interact with each other in a coordinated manner. We will consider two forms of bone formation: (1) endochondral bone formation, which accounts for the majority of bone formation during skeletal growth and (2) pharmacologically induced bone formation, which occurs in adults and is not involved in skeletal growth per se.

Endochondral bone formation is preceded by the formation of cartilage which is then replaced at a later stage by mature bone. Before bone formation takes place, the cartilage undergoes a maturation process involving chondrocyte proliferation and hypertrophy followed by matrix calcification. At this point, the cartilage becomes invaded with blood vessels, the old matrix is degraded and new woven bone is laid down by OBs on the remnants of the calcified cartilage matrix. This woven bone is then resorbed over a period of time and replaced with harder cortical bone by a process of remodeling.

Unlike soft tissues, bone can repair itself in such a way as to be ultimately indistinguishable from the original uninjured tissue and the processes involved in fracture healing recapitulate those seen in embryonic bone formation. The fracture causes mitogens and chemotactic agents to be released from the bone matrix, causing the proliferation and recruitment of precursor cells. These precursor cells then differentiate and synthesize collagen to form a soft cartilaginous granulation tissue which consequently becomes calcified to form a callus. At this point the callus is disorganized, consisting of woven bone but without the mechanical strength of the original bone. In order for this to be achieved, the woven bone is replaced with of new cortical bone by remodeling (Frost 1989). The ability of bones to repair after fracture led to postulation of the existence of "osteogenins," factors stored in the growth plates of bone which direct the formation of new bone (La Croix 1951). It was subsequently shown that the implantation of demineralized bone into rabbit muscle pouches induced the formation of new bone (Urist 1965) and this activity was termed "bone morphogenetic protein" (BMP). The sequence of events seen after the implantation of BMP or demineralized bone matrix shows all of the characteristics normally associated with endochondral bone formation in vivo.

A number of substances are known which stimulate bone formation when administered to adult animals. These include parathyroid hormone (PTH), prostaglandin E_2 (PGE$_2$) and calcitriol (Hock et al. 1988; Norrdin et al. 1990; Erben et al. 1997); however, despite a great deal of research, and the obvious economic implications, the cellular basis for their actions in vivo is not well understood. Treatment with any of these three drugs initially leads to an increase in the number of active OBs on the bone surface (Fig. 1). This is followed by the synthesis of collagen and laying down of osteoid which subsequently becomes mineralized. This mineralized osteoid consists of woven bone and is of relatively low mechanical strength but is eventually remodeled and replaced with lamellar bone. It is largely accepted that the increase in bone formation induced by these drugs is mediated by an increase in OB numbers. The origin of these new OBs is, nonetheless, a source of contention. There are, at present, two predominant hypotheses: (1) activation of bone lining cells and (2) recruitment and proliferation of bone marrow stromal cells. There is evidence for and against both of these hypotheses. The rapidity of the immediate response and the lack of uptake of ^{3}H-thymidine by bone marrow stromal cells (BMSCs) or OBs after treatment with PTH points to the activation of bone lining cells. However, the number of activated OBs often greatly outnumbers the original numbers of bone lining cells, suggesting a proliferative effect and it has been noted that PGE_2 is more effective at sites immediately adjacent to the bone marrow, suggesting the recruitment of BMSCs. These two hypotheses have been hotly debated, although it is probably more realistic to assume that the early events associated with bone anabolism are associated with the differentiation of OB precursors, e.g., bone lining cells, whereas the long-term effects are mediated by the proliferation and recruitment of earlier OB precursors from the bone marrow or periosteum.

Fig. 1a–d. Giemsa stained sections from rats treated with **a** vehicle; **b** 0.5 g/kg calcitriol (3 days after treatment); **c** enlargement of **b** and **d** 0.5 /kg calcitriol (6 days after treatment)

Fig. 3a–d. Photomicrographs of bone marrow-derived colonies stained for **a** total colonies with methylene blue, **b** alkaline phosphatase, **c** collagen with sirius red and **d** immunohistochemically for osteocalcin. *Text see p. 15*

Fig. 15. In situ hybridization analysis of EP4 expression in sections of rat tibia. *Text see p. 36, 37*

Fig. 1a–d.
Legend see p. 4

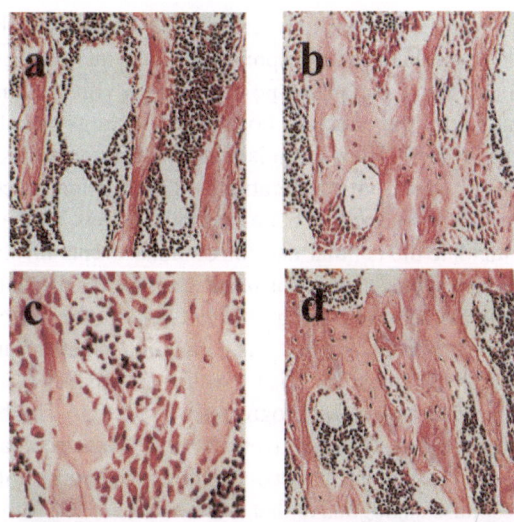

Fig. 3a–d.
Legend see p. 4

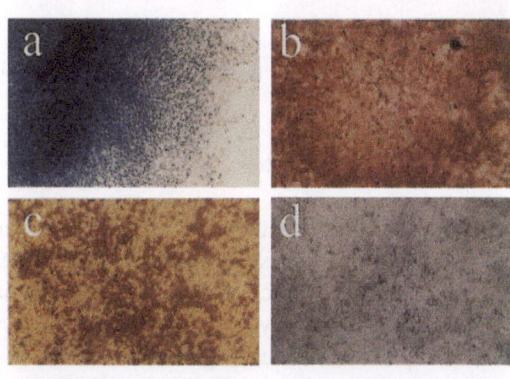

Fig. 15.
Legend see p. 4

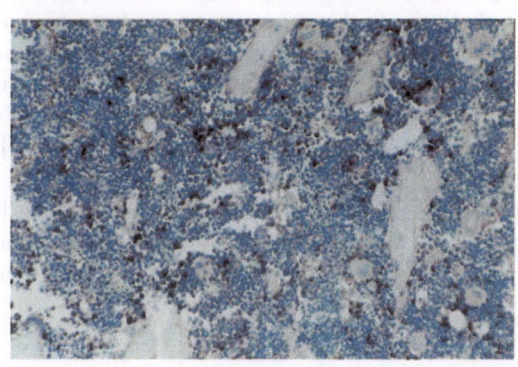

Although it has not yet been established whether either or both of these mechanisms is responsible for the increase in OB numbers, it does mean that a major end-point in any in vitro system purporting to be a model for the effects of bone anabolic drugs must be an increase in OB cell number. The mechanism for this has not yet been identified but may be an effect on proliferation, differentiation or cell death. The main reason for studying bone anabolic effects at the cellular level is to elucidate the cellular and intracellular events involved in the bone anabolic response; indeed, using the techniques presently available, this can only be achieved using cell cultures. Unfortunately, these techniques also require a homogeneous population of cells. This becomes problematic as bone formation is by nature the product of a number of different cell types and, as we progress from cultures of high complexity, such as organ cultures, to relatively homogeneous populations of cells, as in clonal cell lines, it becomes more and more difficult to produce a response analogous to that seen in vivo.

1.3 Organ Cultures: Models of High Complexity

1.3.1 Differentiation of Osteoprogenitor Cells in Folded Chick Periostea

Although the differentiation potential of a cell can be studied in cell culture systems, these systems cannot, of course, lead to any conclusion about the kinetic differentiation path of a cell in its normal spatial environment. Preliminary experiments describe a spatial and functional relationship between osteoprecursor cells and OBs in bone formation in a kinetic model involving folded calvarial periostea in vitro (McCulloh et al. 1990). In this model, periostea from 17-day-old chicks are dissected free, folded over and grown on filters in the presence of ascorbate, β-glycerophosphate and dexamethasone for up to 6 days. In this time a mineralized bone tissue is formed which is virtually identical to that seen in vivo. After 6 days all stages of OB development can be seen, ranging from preosteoblasts through to osteocytes. As no OBs are present at the start of the culture period, this is an ideal model for the study of OB differentiation in a realistic three-dimensional matrix. After continuous labeling with ^{3}H-thymidine over 5 days, 30% of the fibroblastoid, i.e., alkaline phosphatase (APase)-negative, cells and over 95% of the APase-positive cells were labeled. A differential transition to labeling of fibroblastoid and osteogenic cells was found, suggesting that there was no significant regeneration of osteogenic cells from fibroblastoid cells. When all the proliferating osteogenic precursor cells were killed by high doses of ^{3}H-thymidine within 1 or 2 days, no OB differentiated and no bone formation took place. This showed that some of the OB for bone formation originate from a small population of cycling precursors. Clusters of ^{3}H-thymidine-labeled cells continued to be found. The cluster size and number of cells per cluster did not change, but there was a 20-fold increase in the number of clusters during the culture. The clusters were APase-positive cells and located close to the bone surface. The other 60% of the OB-precursor cells differentiate without cell division. The distribution of the APase-positive clusters over the bone surface possibly partly contribute toward the morphology of the bone. On the fifth day, the APase clusters were uniform and formed a layer of cells on or close to the bone surface, leading to a uniformly flat bone.

1.3.2 Bone Formation in Embryonic Rat Calvaria

Although the folded periosteum model enables us to study OB differentiation under realistic conditions, it does not respond to treatment with bone anabolic drugs. Once the cultures are started, they differentiate until a bone-like tissue is formed. This process can be inhibited, but as the cultures are already operating under optimal conditions they cannot proceed any faster. One culture system which can respond to bone anabolic drugs is the embryonic rat calvaria. This, along with similar systems, such as neonatal mouse calvaria and chick long bone cultures, has been used with some success to study bone resorption. However, under defined culture conditions these systems also respond to treatment with bone anabolic drugs with an increase in collagen synthesis and/or thymidine incorporation corresponding to the first phase of the bone anabolic response. In doing so, they also begin to demonstrate the complexity of the problems associated with studying bone formation in vitro.

The response to PTH in vivo is not simple; infusion of PTH leads to a catabolic response whereas pulsatile treatment stimulates bone formation. A similar situation also exists in organ cultures; continuous treatment with PTH inhibits both collagen and DNA synthesis whereas when the bones are treated for 24 h with PTH and then transferred to PTH-free medium, DNA and collagen synthesis are both stimulated. The response to PTH is also, at least to some extent, indirect. PTH stimulates the production of insulin-like growth factor (IGF)-I in these cultures and the addition of blocking antibodies to IGF-I abrogates the stimulatory effects of PTH on collagen synthesis (Canalis et al. 1989). PTH, calcitriol and $PGF_{2\alpha}$ also stimulate PGE_2 synthesis in these cultures. As PGE_2 itself stimulates DNA and collagen synthesis, it must also be a possible mediator of their actions. In contrast to PTH, the PGE_2 effect is IGF-I-independent and is only seen in the presence of physiological concentrations of glucocorticoids (Raisz et al. 1993). The role of glucocorticoids is contentious and this laboratory has argued that their presence is the physiological norm and that they should be present in all such culture systems. Consistent with this is the finding that interleukin (IL)-1, which is also anabolic under certain circumstances in vivo (Boyce et al. 1989; Takada et al. 1994), produces a convincing anabolic effect in vitro only in the presence of cortisol. In isolation both IL-1 and cortisol inhibit collagen synthesis in mouse calvariae, whereas in the presence of cortisol, IL-1 stimulated collagen synthesis (Marusic and Raisz 1991). Because of the complex nature of calvaria, global measurements of collagen or DNA synthesis tell us nothing of the nature of the target cells; however, it is relatively easy to examine these bones histologically. Using this approach it has been shown that the cells which respond to PGE_2 with an increase in DNA synthesis are OB precursor cells found in the bone marrow and not the mature OBs.

1.4 Primary Cell Cultures:
Models of Intermediate Complexity

1.4.1 Nodule Formation by Primary Rat Calvarial Cells

A major breakthrough in the study of bone formation in in vitro bone biology was the development of the nodule model by, among others,

Bellows and co-workers (Bellows et al. 1987). In this model, a mixed population of bone-forming cells from 21 day rat calvaria are cultured with fetal calf serum (FCS), ascorbic acid and organic phosphate. After 6 days, the cultures become confluent; the cells are heterogeneous and include many fibroblastic cells as well as some osteoblastic cells reflecting several different stages of differentiation. By day 8, the cells are densely packed in multilayers, and by day 16 they have synthesized a dense ECM and acquired a translucent appearance. From day 9 onwards, areas of increased cell density containing distinctly more polygonal cells are to be seen, and these continue to increase in size and number up to day 21 at which point they are detectable as large (0.5 mm^2) mineralized bone-like nodules. In cross-section, these nodules display many of the characteristics of woven bone in vivo: fibroblastoid cells were present in the outer layer, inside are cuboid OB-like cells, and deeper inside are osteocyte-like cells, completely surrounded by collagenous matrix. The cells of these nodules form type I collagen, osteonectin, osteocalcin and increased APase activity and synthesize hydroxyapatite (Bhargarva et al. 1988). The development of these nodules, therefore, recapitulates the early stages of bone formation seen in embryonic calvaria (Yoon et al. 1987) or the early stages of fracture healing (Frost 1989).

The number of nodules formed in these cultures is linearly proportional to the number of cells plated out and dilution analysis shows that only 0.3% of the calvaria cell populations isolated are osteoprogenitor cells (i.e., nodule forming cells), and that one osteoprogenitor cell develops into one bone nodule without any need for cooperativity with other cells. The number of precursors per calvaria was determined as approximately 90 if the periosteum was removed and 500 per calvaria if the periosteum was left intact, thus confirming that the periosteum is a rich source of progenitor cells (Bellows and Aubin 1989). The addition of dexamethasone increased the number of nodules from 1 in 340 to 1 in 225, suggesting that dexamethasone is required in order for some precursor cells to progress beyond a particular point in their development (Bellows and Aubin 1989). This effect of dexamethasone was explained by the finding that two separate populations of osteoprogenitor cells exist in the rat calvaria cells: APase-positive and APase-negative. The APase-positive cells could form bone in vitro in the absence of added glucocorticoids, whereas the APase-negative cells required the presence of dexamethasone (Turksen and Aubin 1991).

The in vivo results described above suggest a certain amount of plasticity in the differentiation of OBs which can also be seen in vitro. When mineralized cultures were subcultured, the sequence of proliferation and differentiation was reinitiated and cultures continued to form nodules after as many as four passages (Owen et al. 1990). As proliferation is necessary for the formation of new nodules, this suggests that dedifferentiation has taken place and that the cells local environment, i.e., the ECM and the synthesis of positive or negative growth factors, is a key factor in the expression of its phenotype. It was also found that, when dexamethasone was added to the subcultures, the number and size of the nodules was further increased suggesting that cells which are still undifferentiated are present in the original mineralized cultures (Bellows et al. 1990).

The establishment of this system has enabled some of the underlying molecular mechanisms involved in bone formation to be elucidated. The process of nodule formation was found to be comprised of three distinct time periods: (1) a proliferative phase, (2) a period of ECM maturation and (3) mineralization. The proliferative phase lasts for the first 10–12 days and is characterized by the expression of genes associated with cell division, e.g., H4 histone, and cell growth, e.g., c-*fos* and c-*myc*. Also expressed in this phase are a number of genes associated with the development of connective tissue, i.e., type I collagen, transforming growth factor (TGF)-β and fibronectin. At the end of the proliferative phase, the cells cease to divide. The cell growth-associated genes are then rapidly, and most matrix associated genes more gradually, down-regulated. The second phase then takes place over days 12–21 and is characterized by an increase in the expression of APase and matrix Gla protein (MGP). These reach a maximum by about day 20, the TGF-β then returns to its original levels and the expression of MGP remains high. During this period, the matrix is modified in such a way that mineralization can take place and the cultures can progress into the third phase of development. This phase is characterized by the progressive expression of two bone- related genes osteopontin and osteocalcin. These reach significant levels of expression at about day 16 and reach a maximum by day 20. Their expression is followed by matrix mineralization which also begins to increase at day 16 and continues to increase until the cultures are stopped at day 30.

It was found that, during nodule formation, two restriction points exist, at which signals must be received before the sequence can proceed further. The first restriction point occurs when the proliferation is down-regulated and ECM maturation is induced, and the second when mineralization starts. When proliferation was inhibited on the third day by the addition of hydroxyurea, there was an early increase in APase and osteopontin expression, but not of osteocalcin. When this inhibition was delayed until the ninth day, osteocalcin was expressed, suggesting that the presence of collagen and mineralization are necessary for osteocalcin expression. This is consistent with the behavior of ROS 17/2.8 cells, which cease to divide at confluence and the expression of osteocalcin is increased 20-fold. When ascorbic acid was omitted from the cultures, thus preventing matrix development, the cells continued to grow exponentially, osteocalcin expression was not increased and no mineralization took place. Similarly, when β-glycerophosphate was left out preventing mineralization, osteocalcin was also not expressed.

These results led to a model for OB development which states that proliferation, differentiation and ECM synthesis depend on each other. In this model, proliferation, on the one hand, is associated with the synthesis of ECM; on the other hand, the maturation of ECM down-regulates cell proliferation, and the mineralization of ECM down-regulates ECM maturation (for review see Stein et al. 1990).

1.4.1.1 Modulation of Nodule Formation In Vitro

Although the model discussed above appears to accurately describe some of the early events in bone formation, the process seems to be self regulating. All of the growth factors required for the formation of nodules are synthesized endogenously in the culture itself. The addition of exogenous factors normally serves to inhibit nodule formation. Continuous culture in the presence of epidermal growth factor (EGF) or TGF-β results in a dose-related inhibition of the formation of nodules. Forskolin, which induces adenylate cyclase in this system, showed a biphasic effect, inhibiting nodule formation at 10^{-5} M and stimulating at 10^{-9} M (Turksen et al. 1990). Perhaps most importantly, PTH and calcitriol, two of the best characterized bone anabolic agents which might reasonably be expected to stimulate nodule formation, both pro-

duce a significant time- and dose-dependent inhibition of nodule formation. Indeed, when the cells are treated with higher concentrations of
these agents, nodule formation is almost completely prevented (Bellows
et al. 1990; Ishida et al. 1987). A number of substances are known which
can stimulate nodule formation. Continuous addition of dexamethasone
causes a concentration-related stimulation of the number and size of
nodules, stimulating both parameters by about 100%. It is, however,
thought that this is a permissive action and that glucocorticoids are a
requirement for nodule formation in this system. In addition, PGE_2,
which is a potent bone anabolic agent in vivo, did produce a modest
increase in nodule number (Flanagan and Chambers 1992). These data
together suggest that, although this model can be used to observe and
investigate OB differentiation in a realistic setting, it cannot be used to
investigate the actions of bone anabolic drugs. It is likely that this is
because the cellular targets for the bone anabolic drugs, with the exception of PGE_2, are not present in this particular population. It should also
be noted that, as PGE_2 normally has little effect on proliferation, the
mechanism for the PGE_2-induced increase in nodule formation may be
due to the differentiation of previously noncommitted precursor cells.

1.4.2 Bone Marrow Stromal Cells

There is now evidence to suggest that a key cell in the bone anabolic
response is the bone marrow stromal cell (BMSC). Based on observations in vivo, in vitro and combinations of the two, it is now thought that
OBs and other mesenchymal cells are derived from mesenchymal stem
cells present in the bone marrow which are at least closely related to
BMSC (Owen 1985; Friedenstein 1990). When diffusion chambers
loaded with whole bone marrow cells (BMCs) are implanted into host
animals, bone, cartilage and fibrous material is formed (Bab et al. 1984).
When bone marrow is cultured in vitro, colonies of BMSCs are formed
and the single cell which gives rise to each colony is termed a fibroblastic-colony forming unit (CFU-f). When these colonies are expanded in
culture and re-implanted in diffusion chambers, these also form bone,
cartilage and fibrous tissue thus confirming the origin of the tissue as the
BMSC (Bennett et al. 1991).

A number of observations suggest a role for BMSCs in the increase in bone formation seen in vivo after treatment with bone anabolic agents. Jee et al. (1994) observed that PGE_2 is most effective in inducing bone formation where bone marrow is plentiful and suggested that PGE_2 may act by increasing the recruitment of OB precursor cells. Similarly, treatment with fibroblast growth factor-b (bFGF) induced new bone formation only on the endosteal bone surface, with virtually none being seen on the periosteal surface (Mayahara et al. 1993; Nagai et al. 1995; Nakamura et al. 1995). Under normal physiological conditions in vivo, it is thought that CFU-f are nonproliferative (Owen 1985; Friedenstein 1990), however, new evidence suggests that the increase in bone formation seen after ovariectomy (OVX) or treatment with calcitriol, PTH or PGE_2 is associated with an increase in the number of these cells. When rats were treated with these drugs and the bone marrow cultured ex vivo, an increased number of CFU-f were found in the cultures from the treated rats (Nishida et al. 1994; Theuns et al. 1994).

There is also evidence to suggest that BMSCs are involved in the decreased bone formation seen during aging. Cultures of BMCs from aged rats gave rise to significantly fewer fibroblastic colonies than those from younger rats (Scutt et al. 1996; Egrise et al. 1992; Quarto et al. 1995). Similarly, the rapid formation of trabecular bone seen after bone marrow ablation is much reduced in aged rats compared with younger rats (Liang et al. 1992). Quarto et al. (1995) found that, although BMSCs from aged rats gave rise to fewer colonies than BMSCs from younger rats, when cultured under identical conditions, BMSCs from old and young rats proliferated and differentiated in a similar manner. In addition, when BMSCs from young and old rats were loaded into ceramic blocks and transplanted into a common host animal, similar amounts of bone were produced by both populations of cells. This suggests that the cells per se are not defective and it is more likely that an age-related deficit in OB precursor cells (BMSCs) contributes to the decreased bone formation seen in older rats. The cause of this deficit is not known; however, if this were due to the lack of an endogenous factor(s) or to changes in the local environment of the BMSC, this deficit could be reversed. Consistent with this, preliminary results from this laboratory have also shown that OVX or treatment with bone anabolic agents returns the numbers of colonies back to levels comparable with younger rats (Scutt et al. 1996).

1.4.2.1 Primary High Density Cultures
of Bone Marrow Stromal Cells

As in the nodule model, when BMSCs are grown in culture in the presence of dexamethasone, ascorbic acid and β-glycerophosphate (β-GP), the cells differentiate and acquire an OB phenotype expressing APase and osteocalcin and forming a collagenous matrix which calcifies and resembles woven bone (Leboy et al. 1991; Beresford et al. 1992; Bab et al. 1984; Bennett et al. 1991, Satomura and Nagayama 1991). An important feature of these cultures is that, in contrast to rat calvarial cell-derived nodules, they respond positively to treatment with bone anabolic drugs (Scutt and Bertram 1995). Continuous treatment with PGE_2 causes an increase in total cell number, collagen accumulation, absolute APase activity and osteocalcin synthesis. Interestingly, continuous treatment with PTH or calcitriol inhibits all of these parameters in a similar manner to that seen in the nodule model. If, however, the cells are treated with these drugs only for the first 5 days of the culture period, a stimulation is seen. Similarly, a maximal PGE_2 effect can be achieved by treating the cells at the beginning of the culture period for as little as 48 h (Fig. 2). All of these data suggest that the OB precursors present in the bone marrow are involved in the response to bone anabolic agents. Moreover, the responsive cells are only present for the first few days of the culture period and then either die or differentiate to an unresponsive phenotype. This is also supported by the finding that subcultures of BMSC do not respond to PGE_2, PTH or calcitriol.

Although primary high density cultures of BMSCs can respond positively to treatment with bone anabolic drugs, due to the heterogeneous nature of the cell population it is very difficult to interpret many of the global biochemical parameters measured such as those mentioned above. This is because at least three different processes are taking place simultaneously, i.e., recruitment, differentiation and proliferation, and the individual contributions of these cannot be separated. For example, in a mixed population of cells such as those derived from rat calvaria or bone marrow, if APase activity is increased we cannot say if this is due to: (1) an increase in specific APase activity in pre-existing OBs, (2) an increase in the proliferation of APase positive cells or (3) the recruitment

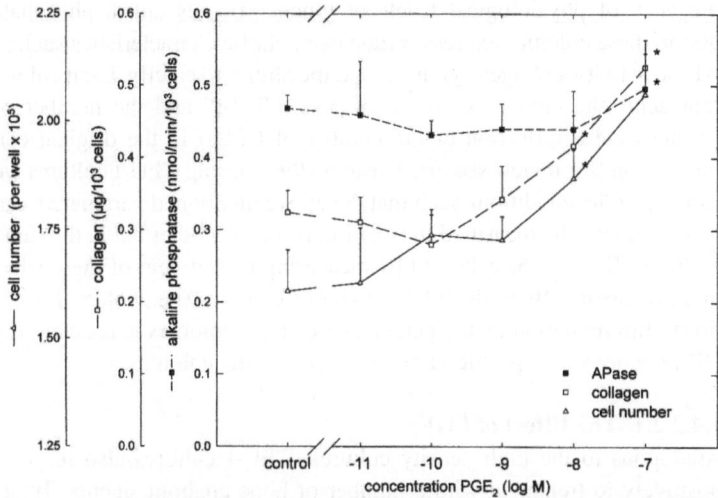

Fig. 2. Effect of prostaglandin (PG)E2 on high density primary BMC cultures. The cells were cultured as described in the text and exposed to various concentrations of PGE2 for the first 5 days of the culture period, after which the medium was changed three times weekly with fresh medium without PGE2. After 14 days, the cultures were stopped and alkaline phosphatase (APase), calcium, collagen and cell number assessed. Results are expressed as means±SD (*n*=8). *Asterisk* denotes a significant difference relative to untreated controls; *p*.05

of uncommitted mesenchymal precursor cells to the osteoblastic lineage. This problem can, to some extent, be avoided by using the CFU-f assay (Fig. 3).

Fig. 3. Color plate, see p. 5

1.4.2.2 Fibroblastic-Colony Forming Unit Cultures
In contrast to high density cultures, in CFU-f cultures, low cell densities are used and the mesenchymal precursor cells adhere and proliferate to form fibroblastic colonies which grow essentially in isolation. In the

presence of physiological levels of glucocorticoids and a phosphate donor, these colonies express certain osteoblastic characteristics such as APase activity, collagen synthesis and the ability to calcify. Each colony represents the clonal expansion of one "CFU-f" and the number of colonies are a reflection of the number of CFU-f in the original cell population (for review see Friedenstein 1990). Using CFU-f cultures we can adjust the conditions such that the above mentioned parameters can to some extent be measured in isolation from each other but in the same cultures. This can be achieved by measuring recruitment of mesenchymal precursor cells as the total number of colonies (i.e., colony formation), differentiation as the percentage of total colonies expressing the OB phenotype and proliferation as the size of the colonies.

1.4.2.2.1 The Effect of PGE$_2$

Analogous to the high density cultures, CFU-f cultures also respond positively to treatment with a number of bone anabolic agents. Treatment with PGE$_2$ causes a caused a concentration-dependent increase in CFU-f formation, the increase normally being of the order of well over 100%. As the cells are cultured in the presence of dexamethasone, ascorbate and β-glycerophosphate, those with osteoblastic potential can differentiate and form a calcified matrix (Fig. 4). A tendency is normally seen towards an increase in the percentage colonies expressing osteogenic markers but this is not reproducible and so it must be concluded that the main effect is on recruitment and not on differentiation. No effect was seen on colony size except at the highest concentration (10^{-7} M), which is, however, comparable to the concentrations used by other authors (Raisz et al. 1990) suggesting a possible direct effect on colony expansion at higher concentrations. These extra colonies can only originate from either nonadherent stromal precursor cells in the culture supernatants or from quiescent stromal precursors attached to the culture vessel. When the nonadherent fraction was cultured in the absence of adherent cells, PGE$_2$ induced the formation of new CFU-f. Furthermore, in an experiment in which the adherent and nonadherent cells were cultivated separately from each other, it was found that PGE$_2$ acts exclusively on the nonadherent fraction, having no effect on the adherent cells (Fig. 5). These data suggest that PGE$_2$ increases recruitment by inducing the transition of nonadherent stromal precursor to an adherent stromal precursor which can then go on to proliferate, form

control

PGE2

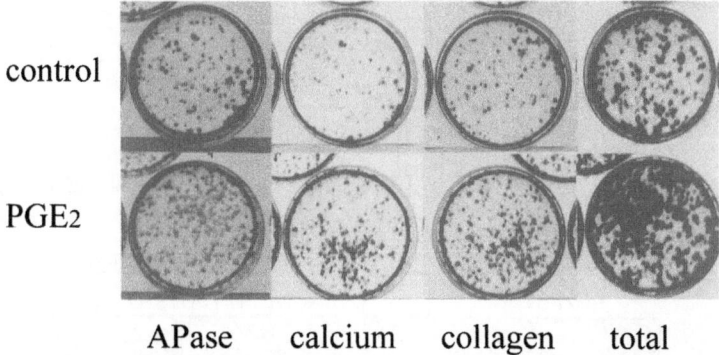

APase   calcium   collagen   total

Fig. 4. Effect of prostaglandin (PG)E2 on colony formation. Bone marrow cells (BMCs) were cultured in the presence or absence of PGE2 (10^{-7} M) as described in the text. After 18 days, the experiment was stopped and the cells stained sequentially for alkaline phosphatase (APase), calcium, collagen and total colonies

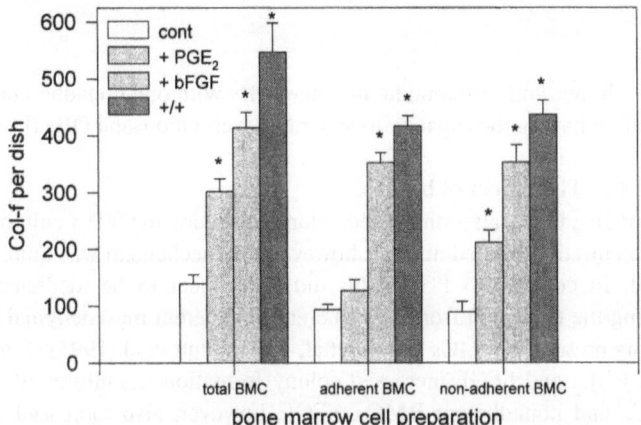

Fig. 5. Do fibroblast growth factor-b (bFGF) and prostaglandin (PG)E2 act on adherent or nonadherent bone marrow cells (BMCs)? Total, adherent and nonadherent BMCs were prepared as described in the text and then challenged for the first 5 days of the culture period with 10^{-7} M PGE2, 10 ng/ml bFGF or both together. After 10 days the cultures were stopped and total number of colonies (Col-f) determined. Results are expressed as means±SD ($n=3$). *Asterisk*, significant difference relative to untreated controls; $p.<0.05$

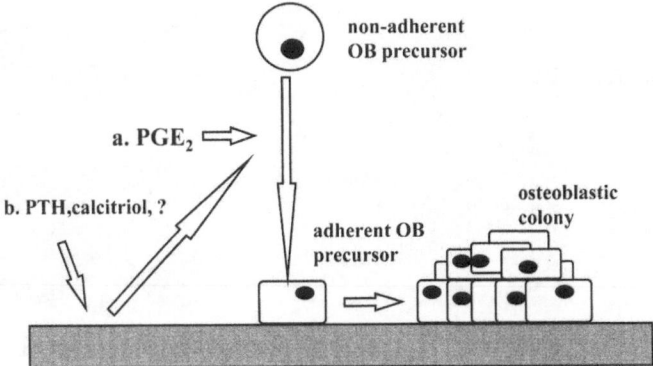

Fig. 6a, b. Possible role of prostaglandin (PG)E$_2$ in osteoblast (OB) recruitment. **a** PGE$_2$ can mediate the transition between nonadherent and adherent OB precursor cells which is then followed by colony expansion. **b** This does not have to be a direct effect, a number of anabolic stimuli are known to cause the local release of PGE$_2$, which could then use the same mechanism to bring about the local recruitment of OB precursor cells

new colonies and differentiate into new OBs with one nonadherent OB precursor having the capacity to generate several thousand OBs (Fig. 6).

1.4.2.2.2 The Effect of bFGF

Like PGE$_2$, bFGF also stimulated colony formation in CFU-f cultures in a concentration-related manner, however, the mechanism was quite different. In contrast to PGE$_2$, this did not appear to be mediated by causing the transition from nonadherent to adherent mesenchymal precursors present in BMCs (Pitaru et al. 1993; Scutt et al. 1995) (Fig. 7). Both PGE$_2$ and bFGF increased colony formation in cultures of total BMCs and nonadherent BMCs. bFGF, however, also increased total colony numbers in adherent BMC cultures in which all nonadherent cells had been removed (Fig. 5). This would suggest that bFGF initiates the proliferation of adherent mesenchymal cells (CFU-f) that under normal conditions would have remained quiescent or died and consequently not have formed colonies. In addition, it was found that addition of bFGF after day 3 of the culture period had no effect on colony formation, suggesting that the bFGF-responsive CFU-f had already

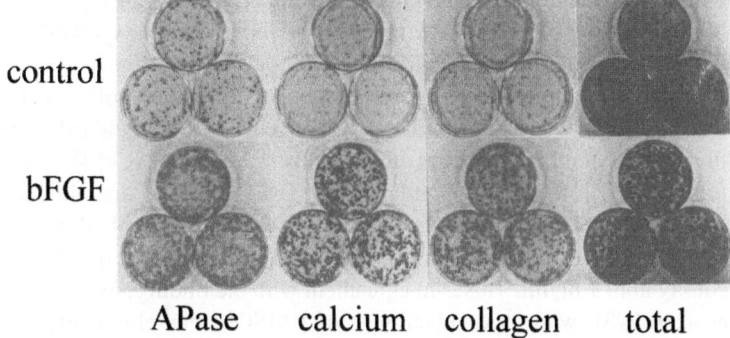

control

bFGF

APase   calcium   collagen   total

Fig. 7. Effect of fibroblast growth factor-b (bFGF) on bone marrow cells (BMCs) fibroblastic colony formation and differentiation. BMCs were cultured as described in the text and exposed to either vehicle or 10 ng/ml bFGF for the first 5 days of the culture period. The medium was then replaced with fresh bFGF-free medium and thereafter changed twice weekly. After 18 days the cultures were stopped, fixed and stained consecutively for alkaline phosphatase (APase), calcium, collagen and total colonies

become apoptotic or unresponsive by this time point (data not shown). Like PGE$_2$, bFGF also stimulated colony formation from nonadherent BMCs which would suggest that bFGF can also induce the transition from nonadherent to adherent mesenchymal precursors. However, as about 20% percent of the nonadherent BMCs spontaneously adhere and form colonies in the absence of PGE$_2$ or bFGF, it is also possible that bFGF is acting through these although it is unlikely that this could account for the observed increase in colony formation.

bFGF stimulates the osteogenic differentiation of colonies. In control cultures, although a significant proportion expressed APase, almost no cultures calcified and only a small number accumulated collagen. The addition of bFGF caused an increase in all types of colony. However, whereas the percentage of AP-positive colonies (Col-AP) remained fairly constant throughout the concentration range in both total BMCs and adherent BMC cultures, the percentage of collagen-positive colonies (Col-ca) and Col-co were increased, until at 10 ng/ml 60%–70% of the colonies were collagen- or calcium-positive. This would suggest then that a fixed percentage of CFU-f have the ability to express APase, which is expressed regardless of the growth factor environment, and that bFGF, or some other growth factor, is required for the synthesis of collagen and subsequent calcification.

bFGF also stimulated the proliferation of individual colonies. By measuring the total numbers of cells in the cultures we could calculate the average number of cells per total number of colonies (Col-f). It was found, in both total BMCs and adherent BMC cultures, that although total colony numbers were increased at concentrations of bFGF as low as 0.1 ng/ml, no significant effect was found on the number of cells per colony until 1 ng/ml. This is in agreement with the findings of Thomson et al. (1993), who used subcultures of BMSCs, thus eliminating the effect of colony formation, and who also found no effect on BMSC cell number until 1 ng/ml. This could mean that colony initiation and proliferation are independent of each other although it is also possible that our method of assessing proliferation was not sensitive enough to detect changes in cell number at low bFGF concentrations. The latter possibility is supported by the findings of Noff et al., who measured ^{3}H-thymidine uptake and found an effect at concentrations as low as 0.01 ng/ml bFGF (Noff et al. 1989).

bFGF also stimulated the amount of collagen accumulated by individual Col-co. Using a similar approach to that used for proliferation, total collagen was calculated and from this the average amount of collagen accumulated per Col-co could be calculated. This has the advantage that only colonies which are synthesizing collagen are included in the calculation, whereas normally results are expressed in terms of total cell number or total protein even though a large and variable number of the cells are not involved in collagen synthesis. As with proliferation, whereas the numbers of Col-co were increased by concentrations of bFGF as low as 0.1 ng/ml, no effect on the levels of collagen per Col-co was seen until 1 ng/ml for total BMCs and 10 ng/ml for adherent BMCs. This would suggest that the process of differentiation is not necessarily linked to the direct control of collagen synthesis. This is supported by the fact that just 5 days exposure to bFGF produced maximal numbers of all types of colony even though significant levels of collagen are not detected in these cultures until at least day 10 (Scutt and Bertram 1995). In addition to these early effects on colony differentiation, long-term exposure to bFGF also increased average colony size and collagen accumulation, suggesting that as well as effects on the precursor cells, bFGF also exerts an direct effect on the more differentiated cells.

1.4.2.2.3 The Role of Glucocorticoids

Glucocorticoids present bone biologists with a particular problem in that: (1) pharmacological doses of glucocorticoids inhibit bone formation in vivo and lead to corticosteroid-induced osteoporosis. (2) Under physiological conditions cortisol is present in the blood at concentrations of up to 5×10^{-7} M. (3) Cell culture studies show that glucocorticoids stimulate OB differentiation and that their presence may be a prerequisite for the differentiation of pre-OBs to OBs (Bellows et al. 1987; Cheng et al. 1994; Rickard et al. 1994). Using CFU-f cultures we can begin to explain this dichotomy.

The PGE_2-induced increase in colony number mentioned above was dependent on the presence of 10^{-7} M dexamethasone which maintained the equilibrium between the nonadherent and adherent stromal precursors (Fig. 8). It was thought possible that the decrease in bone formation caused by pharmacological doses of glucocorticoids may be due to a further inhibition of the nonadherent and adherent transition or due to the inhibition of the action of PGE_2 on this transition. This was found not to be the case as the addition of 10^{-10} or 10^{-9} M dexamethasone had no significant effect on colony numbers, whereas the addition of

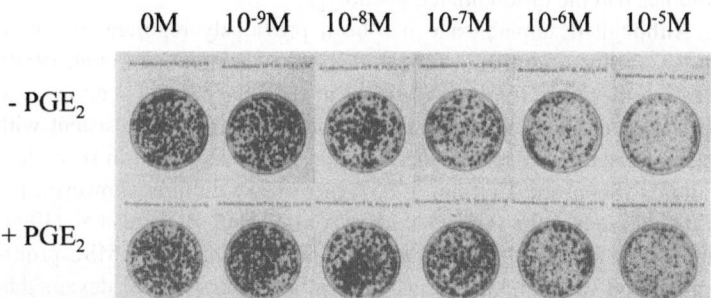

Fig. 8. Effect of dexamethasone and prostaglandin (PG)E$_2$ on fibroblastic-colony forming unit (CFU-f) formation and colony expansion. Total BMCs were exposed to various concentrations of dexamethasone ± PGE$_2$ (10^{-7} M). After 18 days the cultures were stopped, fixed and stained consecutively for alkaline phosphatase (APase), calcium, collagen and total colonies

10^{-8}–10^{-7} M dexamethasone produced an almost maximal decrease in Col-f numbers. The addition of higher concentrations of dexamethasone caused a further small but significant inhibition in Col-f numbers. Similarly, the addition of PGE_2 to control cultures or cultures containing 10^{-10}–10^{-9} M dexamethasone had no effect on CFU-f number, whereas the addition of PGE_2 to cultures containing 10^{-8}–10^{-5} M dexamethasone produced proportionally similar increases in Col-f numbers.

It is well established that dexamethasone stimulates the osteogenic differentiation of human and rat BMSCs. More specifically, dexamethasone has been shown to induce the expression of APase, mineralization, osteocalcin synthesis and PTH-induced cAMP synthesis in monolayer cultures of BMSCs. Similarly, at the mRNA level, mRNAs for APase, osteocalcin, osteopontin and bone sialoprotein were induced after treatment with dexamethasone. Unfortunately, whilst these studies are probably a true reflection of the physiological role of glucocorticoids in OB differentiation, pharmacological concentrations were not tested and so it could not be judged what effect these would have on OB differentiation. In this model, pharmacological concentrations of dexamethasone have only a slight negative influence on osteogenic differentiation by BMSCs. The addition of 10^{-7}–10^{-5} M dexamethasone to CFU-f cultures resulted in a decrease in Col-AP, Col-co and Col-ca numbers relative to cultures treated with 10^{-8} M dexamethasone, but these were still considerably higher than in the control cultures in which almost no colonies had the differentiated phenotype.

Although dexamethasone has been previously reported to inhibit BMSC proliferation (Beresford et al. 1992; Maniatopoulos et al. 1988), these studies were mostly performed in secondary BMSC cultures and with physiological concentrations of dexamethasone. Consistent with this, we have also demonstrated a slight inhibitory effect in secondary cultures (Scutt and Bertram 1995). In primary cultures, however, the situation seems to be somewhat more complicated. Rickard et al. (1994) found no effect of 10^{-8} M dexamethasone on primary rat BMSC proliferation and we found that physiological concentrations of dexamethasone induce an increase in cell number in high density cultures. In the CFU-f cultures, two methods were used to estimate proliferation: (1) determination of total cell number with which average colony cell number was calculated and (2) determination of total colony surface area with which average colony surface area was calculated. Using both of these methods, it was found that although 10^{-8}–10^{-7} M dexamethasone significantly reduced colony numbers (together with total cell number and surface area) there was no dramatic effect on colony size. In contrast, 10^{-6}–10^{-5} M dexamethasone, whilst having an additional inhibitory effect on colony number relative to 10^{-8}–10^{-7} M dexamethasone, had a more significant effect on average colony surface area and cell number, reducing the levels of these dramatically to well below those of the control cultures.

Taken together, these data suggest that physiological levels of dexamethasone (10^{-8}–10^{-7} M): (1) maintain the equilibrium between nonadherent and adherent stromal precursors, allowing PGE_2 to control the recruitment of OB precursors, (2) are a prerequisite for the differentiation of BMSCs to the osteoblastic lineage, and (3) have little effect on BMSC proliferation. Pharmacological doses, by contrast, are much higher and would appear to: (1) have an additional inhibitory effect on BMSC recruitment, i.e., colony formation and, more importantly, (2) drastically reduce BMSC proliferation.

1.4.2.2.4 Analysis of Complex Cell Populations
Using the CFU-f Assay

Although calculating average colony size, either as cell number or as surface area, can give us valuable information regarding the effects of bone active agents, both methods suffer from weaknesses. Firstly, both methods calculate the average colony size and the effects of drugs on individual colonies cannot be assessed. In addition we get no information about the size distribution of the colonies and for this reason we have to treat them as though they have a normal distribution when in fact this may not be the case. Secondly, average colony surface area is only a two-dimensional measurement and gives no information regarding the three-dimensional structure of the colonies. Thirdly, average colony cell number gives an idea of the "colony volume" but, due to the presence of single adherent non-colony forming cells in the cultures, the numbers are always artificially high. In order to rectify this problem we have developed a method to construct three-dimensional "optical density maps" of the cultures with which colony formation can be investigated.

After staining, the CFU-f cultures are photographed using a digital camera and the images transferred to a computer in standard TIFF format. The images are then processed to reduce the background intensity to zero, increase the contrast and remove small variations in intensity and then converted to 8-bit gray scale TIFF images. These are then analyzed using a commercially available image analysis package selecting colonies of at least 20 pixels (corresponding to 1 mm) in diameter and having an intensity of at least 20 gray levels above background. The software then assigns an identity to each colony and calculates its coordinates, surface area and intensity. As the analyzed image has information regarding the staining intensity, a three-dimensional representation of the cultures can also be generated.

Comparisons of colony numbers obtained manually and by image analysis showed a linear relationship and produced a correlation coefficient of 0.99. Similarly, the relationship between colony surface area measured manually and by image analysis was also linear with a correlation-coefficient of $r=0.96$; however, due to the three-dimensional nature of the colonies, this does not necessarily reflect the colony cell number. In order to assess this, known numbers of secondary BMSCs were plated out on petri dishes but in the confines of 5 mm cloning rings. The cells were then allowed to attach overnight and then fixed, stained, and analyzed as described above . The staining index of the colonies was calculated by multiplying the surface area of the colony by the intensity of staining. The staining index also had a linear relationship with cell number up to 5×10^4 cells per colony" at which point the curve leveled off.

Using this methodology, the effects of high concentrations of dexamethasone on colony formation and expansion were re-investigated. BMCs were cultured with differing concentrations of dexamethasone and in the presence or absence of 10^{-6} M PGE$_2$, as described previously. The cultures were then stained with methylene blue and analyzed, first manually as described previously and then by image analysis. The two methodologies produced very similar results with regard to colony numbers. Firstly, there was a dose-related decrease in colony number with increasing concentrations of dexamethasone and, secondly, the addition of PGE$_2$ increased colony numbers but only when dexamethasone was present at concentrations of 10^{-8} M or higher. Anova revealed no significant differences between data acquired using the two different methodologies. The two methods also produced comparable results with regard to colony size, however some significant differences were observed (Fig. 9). Dexamethasone produced a dose-related decrease in

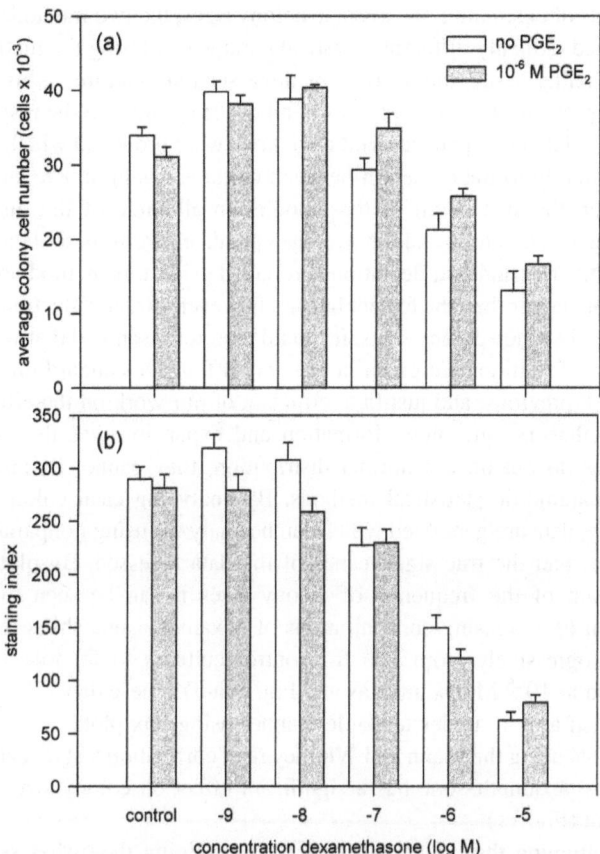

Fig. 9a, b. Effect of dexamethasone and prostaglandin (PG)E_2 on colony expansion. **a** Cultures were treated with various concentrations of dexamethasone ±PGE_2 and then cultured as normal. The cultures were stopped after 18 days and analyzed by **a** calculating total cell number and dividing by the number of colonies or **b** image analysis

colony size of similar magnitude. However, assessing colony size using the methylene blue binding assay revealed an increase in colony size after treatment with 10^{-6} M PGE_2 whereas this was not found when colony size was assessed by image analysis. Due to differences in the methods of calculating the average colony sizes, the two methodologies produced radically different statistical parameters. Using the methylene blue binding assay, the average of three separate cultures is used. As these are themselves the averages of all of the colonies in the respective dishes, relatively small standard deviations were produced which correspond purely to the deviation between cultures. Using image analysis, however, the size of all of the colonies in all three of the dishes is calculated separately, and the average calculated from these data. Consequently, the standard deviations produced using this method are considerably larger than the former but are, however, a truer reflection of the variance between colony sizes. It should therefore be noted that whereas standard deviations are given in Fig. 9a, in Fig. 9b standard errors are given. A previous, and justified, criticism of our work on the effects of dexamethasone on colony formation and expansion was that, as the colonies do not have a normal distribution, they cannot be analyzed using parametric statistical methods. By analyzing each colony individually, data are generated which can be analyzed using nonparametric statistics and the true significance of the data assessed. By plotting a histogram of the frequency of colony sizes it can be seen that the addition of increasing concentrations of dexamethasone shifts the median progressively from 200 for control cultures to 50 for cultures exposed to 10^{-5} M dexamethasone (Fig. 10a–f). These data can also be plotted in a more understandable manner using box plots. Analysis of these data using the Mann and Whitney rank correlation test reveals that, whereas dexamethasone has a significant effect on colony size, PGE_2 does not (Fig. 11).

By aligning the images produced after staining the dishes sequentially for APase, calcium, collagen and total colonies, it is possible to analyze the images and to export the coordinates of each colony into a graphing program such as Sigma Plot. The data can then be plotted in such a way as to show exactly which phenotype each colony is expressing. It can then be shown that some colonies are purely fibroblastic and do not synthesize collagen or express APase. Some colonies express APase and some colonies are collagen-positive whilst some are both APase- and collagen-positive. All colonies which calcify are also APase-positive and collagen-positive.

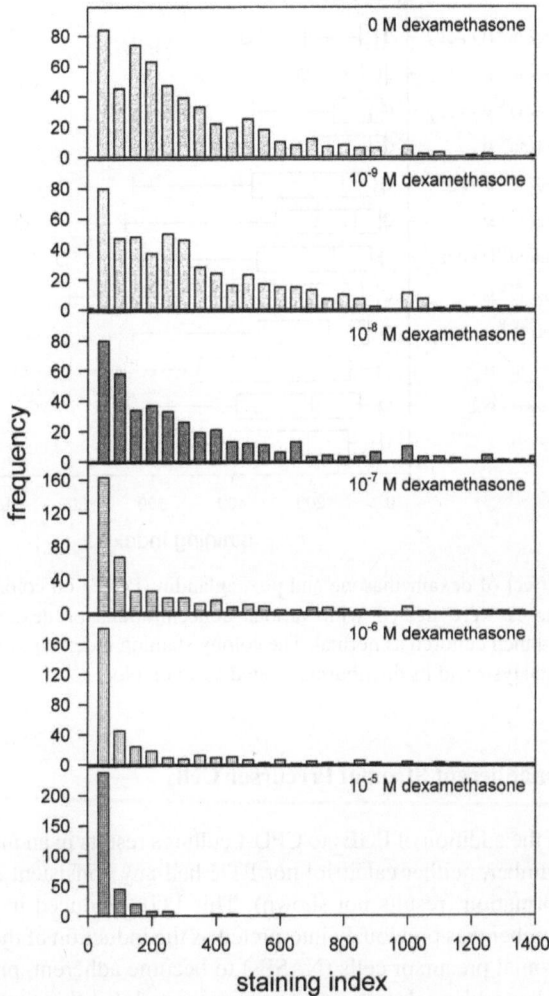

Fig. 10. Effect of dexamethasone on colony expansion. Cultures were treated with various concentrations of dexamethasone and then cultured as normal. The colony staining index was determined by image analysis and its distribution potted as a histogram

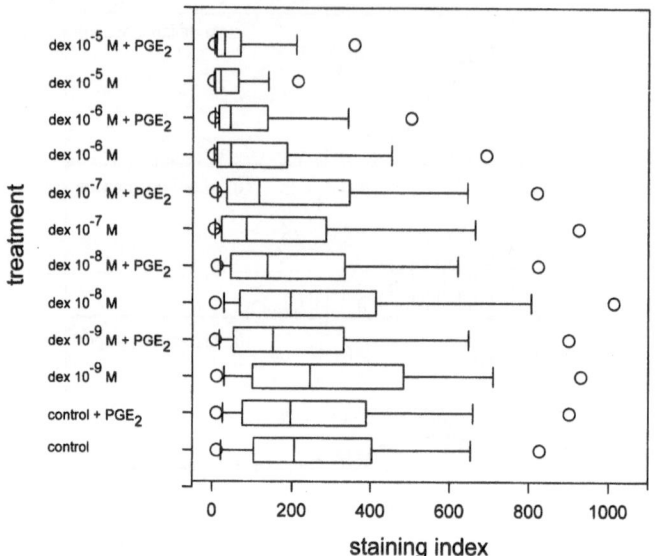

Fig. 11. Effect of dexamethasone and prostaglandin (PG)E$_2$ on colony expansion. Cultures were treated with various concentrations of dexamethasone ±PGE$_2$ and then cultured as normal. The colony staining index was determined by image analysis and its distribution plotted as a box plot

1.4.3 Nonadherent Stromal Precursor Cells

Although the addition of PGE$_2$ to CFU-f cultures results in an increase in colony number, neither calcitriol nor PTH had any consistent effect on colony formation (results not shown). This PGE$_2$-induced increase in colony number was previously interpreted as the induction of the nonadherent stromal precursor cells (NASPs) to become adherent, proliferate and form bone-like colonies and it was suggested that this may explain some of the bone anabolic effects of PGE$_2$. However, this is obviously not the case for calcitriol or PTH and, furthermore, a change in adherence characteristics of the NASP cells would not explain the increase in colony number seen ex vivo after treatment with PTH, calcitriol or PGE$_2$.

The increased number of colonies in ex vivo bone marrow cultures from rats treated with PGE$_2$ (Weinreb et al. 1997), PTH (Nishida et al.

1994) or calcitriol (Erben et al. 1997) suggests an increase in the number of OB precursors. As the NASP cell are possible target precursors, it was decided to investigate the effects of the bone anabolic drugs on NASP cell numbers. As there are at present no molecular markers for the identification of these very early stromal-precursor cells, their numbers can only be assessed using the CFU-f assay. When total BMCs were cultured at high density for 2, 4, 6 or 8 days in the presence or absence of bone anabolic drugs and then the nonadherent cells were transferred to low density petri dish cultures, it was found that those cultures treated with PTH, calcitriol or PGE_2 all developed progressively many more colonies than the control cultures (Fig. 12). These effects appear to be specific and mediated by the relevant receptors, as PTH (3–34), PGB_2 and $24,25(OH)_2$ vitamin D_3 produced no effect at these concentrations.

Although these data strongly suggest that the increase in colony numbers is mediated by a real increase in the number of NASP cells, it could also be due to reduced numbers of NASP cells adhering in the high density cultures. Due to the lack of markers for these cells it is not possible to assess this directly and so pour-off cultures were employed. In these cultures, the nonadherent BMCs are continually poured-off into new petri dishes every 24 h after which fresh medium is added to the adherent cells and these cultures are continued as normal. Due to the short time period between pour-offs, it is assumed that any increase in colony number must be due to either the proliferation of the NASP cells or the differentiation of even earlier precursor cells and cannot involve events taking place at the base of the petri dish. The number of colonies in the control cultures remained constant for at least the first five pour-offs, whereas treatment with PTH, calcitriol or PGE_2 led to a progressive increase in the number of colonies in successive pour-offs. After five pour-offs, treatment with PTH, calcitriol or PGE_2 increased colony numbers by 25%, 55% and 50%, respectively (Fig. 13). If the cumulative numbers of colonies generated are calculated, it can be seen that the control, PTH, calcitriol and PGE_2 treated cultures gave rise to 2830, 3660, 4190 and 3460 colonies, respectively. This continual production of new colonies after each pour-off suggests that under basal conditions new NASP cells are generated and that this can be stimulated by treatment with bone anabolic drugs.

A major problem with cultures of BMCs is their inherent heterogeneity. The vast majority of the cells are of the hematopoietic lineage and

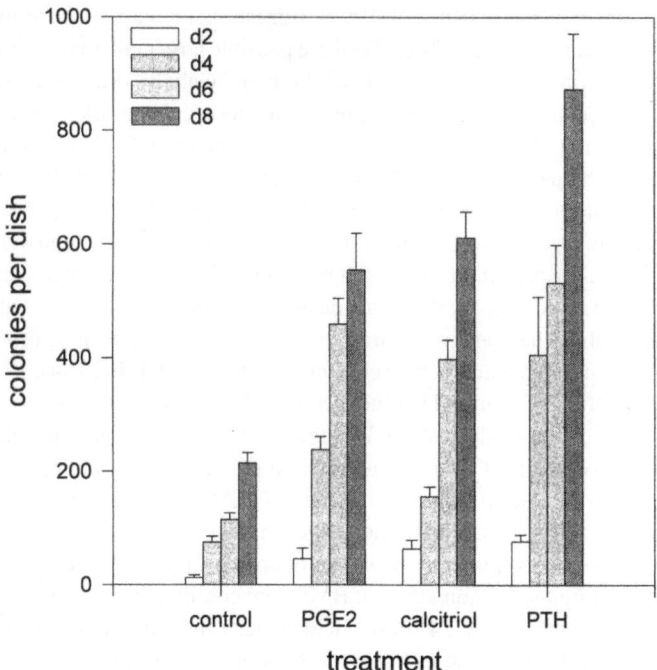

Fig. 12. Effect of bone anabolic drugs on nonadherent stromal precursor cell (NASP) numbers. Some 10^6 bone marrow cells were cultured in 24-well plates for 2, 4, 6 or 8 days. The nonadherent cells were then transferred to petri dishes and the cultures continued as for normal fibroblastic-colony forming unit (CFU-f) cultures. *PTH*, parathyroid hormone

normally only 0.01%–0.1% of the mononuclear cells give rise to fibroblastic colonies. These cells are known as CFU-f and at present no markers are available with which CFU-f can be identified, indeed the very nature of the CFU-f is unclear. It has been suggested that the BMSC are likely to be the CFU-f, however, the data presented here would suggest that the NASP cells are better candidates. Attempts in this laboratory to follow the fate of the NASP cells histochemically, staining for a number of markers, including APase, bromodeoxyuridine uptake, TRAP, ED1, nonspecific esterase, and vimentin, have been without success. The total number of cells transferred at each pour-off decreases

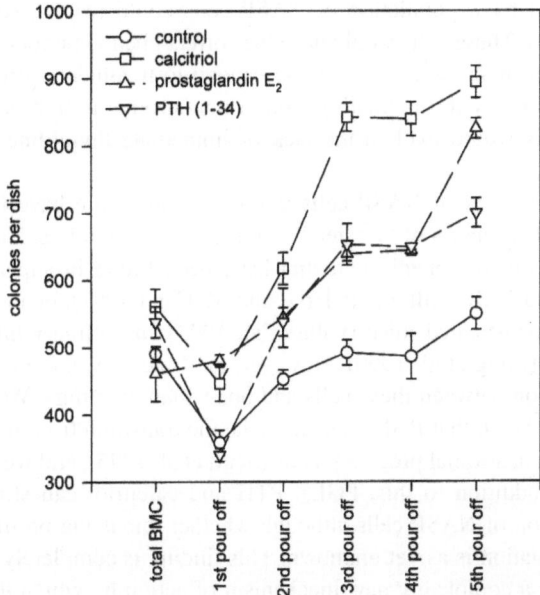

Fig. 13. Effect of bone anabolic drugs on nonadherent stromal precursor cell (NASP) numbers. Some 10^7 bone marrow cells (BMCs) were cultured in petri dishes and, every 24 h, the nonadherent cells were poured-off" into new petri dishes. The poured-off cells were then cultured as normal and fresh medium was added to the adherent cells which were also cultured further. This was continued for up to eight pour-offs. *PTH*, parathyroid hormone

steadily until by the sixth or seventh pour-off it can be calculated that up to 20% of the cells must be NASP cells. Despite this, no markers have been found as yet and we must continue to observe these cells using retrospective assays such as the CFU-f assay.

This lack of markers may be largely due to the fact that most work has concentrated on the fibroblastic BMSCs despite doubts being expressed regarding the role of BMSC in bone formation. These doubts include the inability of fibroblastic cells to cross the epithelial membrane surrounding the bone marrow and the lack of effect of bone active agents on the uptake of radioactive thymidine into BMSCs (Simmons 1995; Dobnig and Turner 1995). The possibility that these effects are mediated by a population of NASP cells could resolve some of the problems. These cells would have the correct phenotype for crossing the epithelial membrane, and if they respond to treatment with bone anabolic drugs by differentiating from a non-stromal to a stromal phenotype, this would explain the lack of immediate thymidine uptake by BMSCs.

The concept of NASP cells is not new and there have been many apocryphal reports of "floaters" in bone marrow cultures. Nonadherent BMCs with osteogenic potential have been found in supernatants of long-term BMC cultures (Clarke and McCann 1991)obtained from 5-fluorouracil-treated mice (Falla et al. 1993) and primary human bone marrow (Long et al. 1990). However, until recently, there has been no connection between these cells and bone anabolic drugs. We have previously shown that PGE_2 can stimulate the transition from nonadherent to adherent stromal precursor cells (Scutt et al. 1995), and we now show that, in addition to this, PGE_2, PTH and calcitriol can stimulate the production of NASP cells although whether this is via proliferation or differentiation is as yet unknown. This finding is completely novel and suggests a completely new mechanism of action by which these drugs may be acting. The nonadhesivness of these cells in vitro is not necessarily a reflection of the in vivo situation. The nonadherent and adherent phenotypes may in fact reflect the different adhesion characteristics necessary for existence either in the bone marrow space or on the bone surface. The physiological relevance of this finding can, of course, only be established in vivo; however, until the cell biology of these cells is better understood and methodologies developed so that they can be identified in situ, this will not be possible.

1.5 Osteoblastic Cell Lines: Models of Low Complexity

It is quite evident that, although the above described models can to some extent recapitulate bone formation in vitro and in some cases respond to bone anabolic drugs in a manner similar to that seen in vivo, they can provide little evidence regarding the cellular and intracellular events involved in these processes. Some limited information can be obtained by the use of specific agonists or antagonists; however, due to the heterogeneous nature of the cultures, this information cannot be localized to a particular cell type. This kind of information can only be realistically obtained using clonal cell lines and, at this point, interpretation of the data begins to become difficult as the cultures cease to respond to bone anabolic drugs in a manner similar to that seen in vivo. PTH, calcitriol and PGE_2 all tend to inhibit OB proliferation and collagen synthesis in clonal OB cell lines although under certain conditions, and with a number of cell lines, a stimulation can be obtained. Another major criticism is that the cell lines lack many of the typical osteoblastic characteristics.

In recent years considerable progress has been made in the development of immortalized cells in culture in order to obtain cell systems that proliferate rapidly, have an extended life span and maintain a normal phenotypic state. Normal mammalian cells have a proliferative life span in culture which is curtailed by senescence, an apparently predetermined growth arrest (Hayflick 1965). Human cells rarely immortalize spontaneously in vitro (Macieira-Coelho 1988); the rare and unpredictable nature of this event precludes attempts at reproducible immortalization of the discrete cell lineages found within bone. Expression of SV40 T-antigen in human cells extends the proliferative life span for 20–30 population doublings beyond the point of senescence (Ide et al. 1984). Immortalization is rarely absolute and generally this extended growth period culminates in crisis, during which cells cease to divide, apparently through mitotic failure, and eventually begin to die. Current methods to immortalize cells can delay but not prevent senescence.

The immortalized state can be generated by several specific viral oncogenes, e.g., the large T-antigen genes of viruses like SV40 and polyoma, which appear to increase the immortalization frequency of human cells (Jat and Sharp 1986). The preferred immortalizing technique has been stable transfection of the SV40 large T-antigen gene into target cells, although genes encoding the nuclear proteins v-Myc and v-H-Ras have also been employed. In addition the adenovirus E1A and E1B promoter regions and the human papilloma viral 16 (HPV16) E7 and E6 proteins have been used as immortalizing agents. The SV40 T-antigen is currently the most convenient because it has been the most extensively characterized and vectors with the gene are currently available (Spelsberg et al. 1995). Constitutive expression of the SV40 T-antigen can be circumvented with the use of naturally occurring temperature-sensitive mutants (SV40tsTag). These proteins have single amino acid changes which make them thermolabile. At 39°C (the nonpermissive temperature) the tertiary structure of the protein is altered resulting in a loss of its ability to bind to p53 and retinoblastoma (Rb) proteins, whereas at 33°C (the permissive temperature) SV40tsTag is correctly folded and sequesters p53 and Rb. This regulation of the immortalization process can yield conditionally immortalized cells that express an immortalized state at 33°C and express a nonimmortalized state at 39°C. Temperature-sensitive mutants have been used to conditionally immortalize human adult (HOB-02-C1) (Bodine et al. 1996a) and fetal OBic cells (hFOB) (Harris et al. 1995), a pre-osteocytic cell line from adult human bone (HOB-01-C1) (Bodine et al. 1996b) and a bipotential osteoprogenitor cell line from human bone marrow (Houghton et al. 1998).

Our laboratories have immortalized human bone marrow stromal cells (hBMSC) that were isolated from the ribs of patients undergoing cardiothoracic surgery. The hBMSC were immortalized by infection with an amphotropic retrovirus transducing a temperature sensitive mutant of SV40 large T-antigen. The helper-virus-free clone PA/tsA58-U19/8 has been shown to package a full length large T gene sequence retaining both tsA58 and U19 mutations, plus the neo^R selectable marker gene (Stamps et al. 1994). The neo^R allows subsequent selection of antibiotic-resistant colonies using neomycin/G418. Using retroviral vectors, which have become widely used for gene transfer and expression of heterologous proteins, problems of toxicity and inefficient gene transfer can be overcome. Their small genome (pproxequal9 kb) also makes them easy to manipulate and use. Retroviruses are taken up by the target cells via specific cell-surface receptors and, consequently, the efficiency of gene transfer is high, with frequencies of approximately 1% being achieved. The limitation in their use is that retroviral vectors can only infect cells that are actively dividing and not those that are terminally differentiated. The adenoviruses are a family of oncogenic DNA viruses that are able to infect fully differentiated (nondividing) cells. Recently adenovirus-*ori*–SV40*ts*A 209, which encodes a temperature-sensitive large T-antigen mutant, has been used to establish osteoblastic (Bodine et al. 1996a) and pre-osteocytic (Bodine et al. 1996b) cell lines from adult human bone. Both of these cell lines are derived from primary cell phenotypes that are thought to be fully differentiated and thus resistant to retroviral transduction. No selectable marker is included in this vector, which may cause isolation of cells expressing the oncogene problematic.

A major advantage of these techniques is that they offer the potential to obtain continually phenotypically consistent and stable cell populations large enough for biochemical analysis. One caveat is that the phenotype expressed by cells in long-term cultures may reflect adaptations to the in vitro environment as well as their intrinsic in vivo phenotype. It is possible that immortalization may affect cell behavior, both growth and phenotypic expression. Cloning offers the advantage that the phenotype of a cell population, even following extended periods in culture, can be traced back to a single progenitor cell. However clonally derived cells are phenotypically heterogeneous and may undergo phenotypic drift with time, which demonstrates the need for periodic phenotypic checks.

1.6 In Situ Hybridization: A Unifying Methodology?

In order to interpret data obtained from the models described above at all levels of complexity, there is a need for a common method of analysis which can be applied to all of the models and the data compared. One possible methodology may be in situ hybridization (ISH). ISH is a powerful technique which allows specific nucleic acid sequences to be detected in morphologically preserved tissue sections or cells. ISH depends on the hybridization of a labeled nucleic acid probe to a complementary sequence of mRNA. The technique provides anatomical evidence as to which cells in a tissue or organ are transcribing which genes and can be used in combination with immunohistochemistry to examine and compare gene expression at both mRNA and protein levels.

The technique was originally described in 1969 by Gall and Pardue (1969). A variety of probes, from double-stranded DNA probes, single-stranded RNA probes or single-stranded DNA oligonucleotide probes, can be used, each with distinct advantages and disadvantages. ISH was initially performed using radioactive probes. The principle disadvantage of these radioactive systems include the need to work with radioisotopes, the time taken for exposure and, more importantly, poor localization within tissue. Nowadays, nonradioactive labels such as digoxigenin and biotin are used. Furthermore, different haptens can be used to label different probes to provide information on the expression of different genes within the same tissue. The principle advantages of nonisotopic labeling systems are the ease of use and the speed of detection, and because the system is based on the methods of immunohistochemistry, precise intercellular localization, and even intracellular localization, can be obtained.

ISH can therefore be used to obtain information at all levels of complexity and we have used this approach to begin to establish a cellular basis for the actions of PGE_2 on BMCs. We have previously shown that one action of PGE_2 is to induce the transition from NASP cells to adherent stromal precursors (Scutt and Bertram 1995). Furthermore, using pharmacological methodology, we could also show that this was mediated via an increase in intracellular cAMP via either the EP_2 or EP_4 PGE_2 receptor (Scutt et al. 1995). Unfortunately, due to the limitations of working with complex cell populations, we can obtain no more information regarding the nature of the cells involved. Using PCR analysis, we can show that the relevant receptors are indeed expressed in primary BMC cultures but this technique can tell us nothing regarding the cellular distribution; and at present we cannot enrich or purify the cells further (Fig. 14). By using the same probes to analyze either cytospin preparations of primary BMCs or sections of rat tibia, it can be seen that the EP_4 receptor can be localized to a discrete and relatively

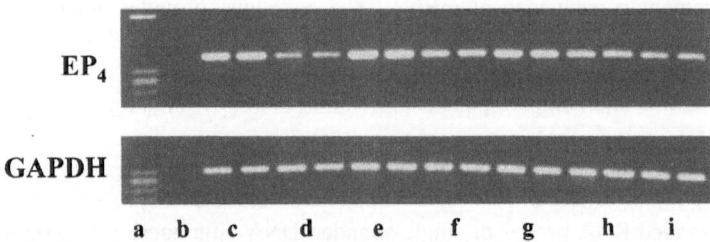

Fig. 14 a–i. RTPCR analysis of EP_4 receptor expression in bone marrow cells. **a** DNA standard; **b** no cDNA; **c** spleen; **d** kidney; **e** primary bone marrow stromal cells (BMSCs); **f** primary BMSC plus dexamethasone; **g** adherent BMSCs; **h** adherent BMSCs plus dexamethasone; **i** adherent BMSCs plus dexamethasone + prostaglandin (PG)E_2

small subpopulation of BMCs (Fig. 15). Although we are not yet in a position to unequivocally identify the osteoblastic precursor cells present in the bone marrow, we have considerably reduced the number of cell that must be considered.

Fig. 15. Color plate, see p. 5

References

Bab I, Howlett CR, Ashton BA, Owen ME (1984) Ultrastructure of bone and cartilage formed in vivo in diffusion chambers: Clin Orthop 187:243–54

Bellows CG, Aubin JE (1989) Determination of numbers of osteoprogenitors present in isolated fetal rat calvaria cells in vitro. Dev Biol 133:8–13 1989

Bellows CG, Heersche JN, Aubin JE (1990) Determination of the capacity for proliferation and differentiation of osteoprogenitor cells in the presence and absence of dexamethasone. Dev Biol 140:132–8

Bellows CG, Ishida H, Aubin JE, Heersche JN (1990) Parathyroid hormone reversibly suppresses the differentiation of osteoprogenitor cells into functional osteoblasts. Endocrinology 127:3111–6

Bellows CJ, Aubin JE, Heersche JNM (1987) Physiological concentrations of glucocorticoids stimulate formation of bone nodules from isolated rat calvarial cells in vitro. Endocrinology 121:1985–1992

Bennett JH, Joyner CJ, Triffitt JT, Owen ME (1991) Adipocytic cells cultured from marrow have osteogenic potential. J Cell Sci 99:131–9

Beresford JN, Bennett JH, Devlin C, Leboy PS, Owen ME (1992) Evidence for an inverse relationship between the differentiation of adipocytic and osteogenic cells in rat marrow stromal cell cultures. J Cell Sci 102:341–351

Bhargava U, Bar-Lev M, Bellows CG, Aubin JE (1988) Ultrastructural analysis of bone nodules formed in vitro by isolated fetal rat calvaria cells. Bone 9:155–63

Bodine PV, Trailsmith M, Komm BS (1996a) Development and characterization of a conditionally transformed adult human osteoblastic cell line. J Bone Miner Res 11:806–819s

Bodine PV, Vernon SK, Komm BS (1996b) Establishment and hormonal regulation of a conditionally transformed preosteocytic cell line from adult human bone. Endocrinology, 137:4592–604

Boyce BF, Aufdemorte TB, Garrett IR, Yates AJP (1989) Mundy, GR Effects of interleukin-1 on bone turnover in normal mice. Endocrinology 125:1142–1150

Canalis E, Centrella M, Burch W, McCarthy T (1989) Insulin-like growth factor I mediates selective anabolic effects of parathyroid hormone in bone cultures. J Clin Invest 83:60–65

Cheng SL, Yang JW, Rifas L, Zhang SF, Avioli LV (1994) Differentiation of human bone marrow osteogenic stromal cells in vitro: induction of the osteoblast phenotype by dexamethasone. Endocrinology 134:277–286

Clarke E, McCann SR, (1991) Stromal colonies can be grown from Nonadherent cells in human long-term bone marrow cultures. Eur J Haematol 46:296–300

Dobnig H, Turner RT (1995) Evidence that intermittent treatment with parathyroid hormone increases bone formation in adult rats by activation of bone lining cells. Endocrinology 122:1146–1150

Egrise D, Martin D, Vienne A, Neve P, Schoutens (1992) The number of fibroblastic colonies formed from bone marrow is decreased and the in vitro proliferation rate of trabecular bone cells increased in aged rats. Bone 13:355–61

Erben RG, Scutt A, Miao D, Kollenkirchen U, Haberey M (1997) Short-term treatment of rats with high-dose calcitriol stimulates bone formation in vivo and increases the numbers of osteoblast precursor cells in the bone marrow. Endocrinology 138: 4629–4635

Falla N, Van Vlasselar P, Bierkens J et al. (1993) Characterisation of a 5-fluorouracil-enriched osteoprogenitor population of the murine bone marrow. Blood 82:3580–3591

Flanagan AM, Chambers TJ (1992) Stimulation of bone nodule formation in vitro by prostaglandins E1 and E2. Endocrinology 130:443–8

Friedenstein AJ (1990) Osteogenic stem cells in the bone marrow. J Bone Miner Res 7:243–272

Frost HM (1989) The biology of fracture healing. An overview for clinicians. Part I. Clin Orthop 248:283–93

Gall JG, Pardue ML (1969) Formation and detection of RNA-DNA hybrid molecules in cytological preparations. Proc Natl Acad Sci USA 63: 378–383

Harris SA, Enger RJ, Riggs BL (1995) Development and characterization of a conditionally immortalized human fetal osteoblastic cell line. J Bone Miner Res 10:178–186

Hayflick L (1965) The limited in vitro lifetime of human diploid cell strains. Exp Cell Res 37:614–636

Hock JM, Gera I, Fonseca J, Raisz LG (1988) Human parathyroid hormone (1–34) increases bone mass in ovariectomized and orchidectomized rats. Endocrinology 122:2899–2904

Houghton A, Oyajobi BO, Foster GA, Russell RGG, Stringer BMJ (1998) Immortalization of human marrow stromal cells by retroviral transduction with a temperature-sensitive oncogene: identification of bipotential precursor cells capable of directed differentiation to either an osteoblast or adipocyte phenotype. Bone 22: 7–16s

Ide T, Tsuji Y, Nakashima T, Ishibashi S (1984) Progress of aging in human diploid cells transformed with a tsA mutant of simian virus 40. Exp Cell Res 150:321–328

Ishida H, Bellows CG, Aubin JE, Heersche JNM (1987) Characterisation of the 1,25-(OH)$_2$D$_3$-induced inhibition of bone nodule formation in long-term cultures of fetal rat calvaria cells. Endocrinology 132:61–66

Jat PS, Sharp PA (1986) Large T antigens of SV40 and Polyomavirus effeciently establish primary fibroblasts. J Virol 59:746–750

Jee WSS, Ma YF, Li M, Liang X, Lin BY, Li XJ, Ke HZ, Mori S, Setterberg RB, Kimmel DB (1994) Sex steroids and prostaglandins in bone metabolism. Ernst Schering Research Foundation workshop 9, Sex steroids and bone, Springer-Verlag, Berlin, Heidelberg, New York, pp 119–150

La Croix P (1951) The organisation of bones. Blakiston, Philadelphia

Leboy PS, Beresford JN, Devlin C, Owen (1991) ME Dexamethasone induction of osteoblast mRNAs in rat marrow stromal cell cultures. J Cell Physiol 146:370–8

Liang CT, Barnes J, Seedor JG, Quartuccio HA, Bolander M, Jeffrey JJ, Rodan GA (1992) Impaired bone activity in aged rats: alterations at the cellular and molecular levels. Bone 13:435–41

Long MW, Williams JL, Mann KG (1990) Expression of human bone-related proteins in the hematopoietic microenvironment. J Clin Invest 86:1387–1395

Macieira-Coelho A (1988) Biology of normal proliferating cells in vitro – relevance for in vivo aging. Interdisciplinary Topics Gerontology 23:1–212

Maniatopoulos C, Sodek J, Melcher AH (1988) Bone formation in vitro by stromal cells obtained form bone marrow of young rats. Cell Tiss Res 254:317–330

Marusic A, Raisz LG (1991) Cortisol modulates the actions of interleukin-1 alpha on bone formation, resorption, and prostaglandin production in cultured mouse parietal bones Endocrinology 129:2699–2706

Mayahara H, Ito T, Nagai H, Miyajima H, Tsukuda R, Taketomi S, Mizoguchi J, Kato K (1993) In vivo stimulation of endosteal bone formation by basic fibroblast growth factor in rats. Growth Factors 9:73–80

McCulloch CA, Fair CA, Tenenbaum HC, Limeback H, Homareau R (1990) Clonal distribution of osteoprogenitor cells in cultured chick periostea: functional relationship to bone formation. Dev Biol 140:352–61

Nagai H, Tsukuda R, Mayahara H (1995) Effects of basic fibroblast growth factor (bFGF) on bone formation in growing rats. Bone 16:367–73

Nakamura T, Hanada K, Tamura M, Shibanushi T, Nigi H, Tagawa M, Fuku-
   moto S, Matsumoto T (1995) Stimulation of endosteal bone formation by
   systemic injections of recombinant basic fibroblast growth factor in rats.
   Endocrinology 136:1276–84
Nishida S, Yamaguchi A, Tanizawa T, Endo N, Mashiba T, Uchiyama Y, Suda
   T, Yoshiki S, Takahashi HE (1994) Increased bone formation by intermittent
   parathyroid hormone administration is due to the stimulation of proilifera-
   tion and differentiation of osteoprogenitor cells in the bone marrow. Bone
   15:717–723
Noff D, Pitaru S, Savion N (1989) Basic fibroblast growth factor enhances the
   capacity of bone marrow cells to form bone-like nodules in vitro. FEBS
   Lett 250:619–621
Norrdin RW, Jee WSS, High WB (1990) The role of prostaglandins in bone in
   vivo. Prostaglandins Leukot Essent Fatty Acids 41:139–145
Owen M (1985) Lineage of osteogenic cells and their relationship to the stromal system. In: Peck WA (ed) Bone and mineral research, vol. 3, Elsevier,
   Amsterdam, pp 1–25
Owen TA, Aronow M, Shalhoub V, Barone LM, Wilming L, Tassinari MS,
   Kennedy MB, Pockwinse S,Lian JB, Stein GS (1990) Progressive development of the rat osteoblast phenotype in vitro: reciprocal relationships in expression of genes associated with osteoblast proliferation and differentiation
   during formation of the bone extracellular matrix. J Cell Physiol
   143:420–30
Pitaru S, Kotov-Emeth S, Noff D, Kaffuler S, Savion N (1993) Effect of basic
   fibroblast growth factor on the growth and differentiation of adult stromal
   bone marrow cells: enhanced development of mineralized bone-like tissue
   in culture. J Bone Miner Res 8:919–929
Quarto R, Thomas D, Liang CT (1995) Bone progenitor cell deficits and the
   age-associated decline in bone repair capacity. Calcif Tissue Int 56:123–9
Raisz LG, Fall PM (1990) Biphasic effects of prostaglandin E2 on bone formation in cultured fetal rat calvariae: interaction with cortisol. Endocrinology
   126:1654–1659
Raisz LG, Fall PM, Gabbitas BY, McCarthy TL, Kream BE, Canalis E (1993)
   Effects of prostaglandin E_2 on bone formation in cultured fetal rar calvariae: Role of insulin-like growth factor-I. Endocrinology 133:1504–1514
Rickard DJ, Sullivan TA, Shenker BJ, Leboy PS, Kazhdan I (1994) Induction
   of rapid osteoblast differentiation in rat bone marrow stromal cell cultures
   by dexamethasone and BMP-2. Dev Biol 161:218–228
Satomura K, Nagayama M (1991) Ultrastructure of mineralized nodules
   formed in rat bone marrow stromal cell culture in vitro. Acta Anat
   142:97–104

Scutt A, Bertram P (1995) Bone marrow cells are targets for the anabolic actions of PGE$_2$: Induction of a transition from nonadherent to adherent osteoblast precursors. J Bone Miner Res 10:474–487

Scutt A, Kollenkirchen U, Bertram P (1996) The effect of age and ovariectomy on fibroblastic colony-forming unit numbers in rat bone marrow. Calc Tiss Int 59:309–310

Scutt A, Zeschnick M, Bertram P (1995) PGE$_2$ induces the transition from nonadherent to adherent bone marrow mesenchymal precursor cells via a cAMP/EP$_2$ mediated mechanism. Prostaglandins 49:383–395

Simmons DJ (1995) The in vivo role of bone marrow fibroblast-like stromal cells. Calc Tissue Int 58:129132

Spelsberg TC, Harris SA, Riggs BL (1995) Immortalized osteoblast cell systems (new human fetal osteoblast systems). Calcif Tissue Int (Suppl 1) 56: S18–S21

Stamps AC, Davies SC, Burman J, OHare MJ (1994) Analysis of proviral integration in human mammary epithelial cell lines immortalized by retroviral infection with a temperature-sensitive SV40 T-antigen construct. Int J Cancer 57:865–874.

Stein GS, Lian JB, Owen TA (1990) Relationship of cell growth to the regulation of tissue-specific gene expression during osteoblast differentiation. FASEB J 4:3111–23

Takada M, Yamamoto R, Morita R (1994) Chronic intramedullary infusion of interleukin-1 alpha increases bone mineral content in rats. Calcif Tiss Int 55:103–108

Theuns HM, McOsker JE, Offerman E, DSouza SM (1994) Effects of parathyroid hormone and prostaglandin E2 on bone cell proliferation and differentiation in aged rats. J Bone Miner Res (Suppl 1) 9: S395

Thomson BM, Bennett J, Dean V, Triffitt J, Meikle MC, Loveridge N (1993) Preliminary characterisation of porcine bone marrow stromal cells: skeletogenic potential, colony-forming activity, and response to dexamethasone, transforming growth factor β, and basic fibroblast growth factor. J Bone Miner Res 10:1173–1183

Turksen K, Aubin JE (1991) Positive and negative immunoselection for enrichment of two classes of osteoprogenitor cells. J Cell Biol 114:373–84

Turksen K, Grigoriadis AE, Heersche JN, Aubin JE (1990) Forskolin has biphasic effects on osteoprogenitor cell differentiation in vitro. J Cell Physiol 142:61–9

Urist M R (1965) Bone formation by autoinduction. Science 150:893–9

Weinreb M, Suponitzky I, Keila S (1997) Systemic administration of an anabolic dose of PGE$_2$ in young rats increases the osteogenic capacity of bone marrow. Bone 20:521–526

Yoon K, Buenaga R, Rodan GA (1987) Tissue specificity and developmental expression of rat osteopontin. Biochem Biophys Res Commun 148:1129–36

2 The Identification and Isolation of Cells of the Osteoblast Lineage in Cultures of Adult Human Bone-Derived Cells by Dual Labeling with the Monoclonal Antibodies STRO-1 and B4-78

K. Stewart, C. Jefferiss, J. Screen, S. Walsh, and J.N. Beresford

2.1 Introduction

The cellular etiology of bone loss in osteoporosis is complex. At the tissue level it results from an imbalance between the activities of osteoclasts and osteoblasts such that, at the completion of each remodeling cycle, the amount of bone that remains is less than that originally present. The magnitude of the bone deficit and the precise mechanism responsible for its occurrence vary according to the underlying cause of the disease and its stage of progression. In the postmenopausal variant there is evidence that bone loss during the immediate perimenopausal period results from an increase in the number, activity and, possibly, functional life span of osteoclasts that is not matched by similar changes in the osteoblast population. During the later stages of the disease, however, there is evidence that the continuing loss of trabecular bone results from a decrease in bone formation that is due to a reduction in the number of osteoblasts. It has been proposed that this same mechanism is a major cause of bone loss in all other forms of osteoporosis (Parfitt 1990a, b, 1992; Cohen-Solal et al. 1991; Compston et al. 1989; Jilka et al. 1996).

By assimilating evidence from a variety of sources it is possible to postulate three cellular mechanisms that could, alone or in combination, account for a reduction in the number of osteoblasts:

1. Insufficient recruitment of cells to the osteogenic pathway of differentiation
2. Impaired proliferation and/or premature differentiation of osteoblast precursors
3. Premature (programmed) cell death

Irrespective of the precise cause(s) of the osteoblast deficit, our ability to prevent and, most importantly, treat the clinical manifestations of osteoporosis will depend ultimately on a thorough understanding of the regulation of osteoblast development in the adult human skeleton.

One potential source of osteogenic precursors in the adult skeleton is the marrow stroma. It has long been recognized that, in addition to providing structural and functional support for hematopoiesis, this tissue contains primitive precursors (CFU-F) capable of extensive proliferation and of giving rise to cells of multiple marrow stromal cell lineages, including osteoblasts (Beresford 1989). The precise number and hierarchy of marrow stromal cell lineages is currently unknown. The model proposed originally by Owen and Friedenstein envisaged the existence of a population of multipotential stem cells, capable of self renewal and of giving rise to committed progenitors for each of the marrow stromal cell lineages (Owen 1985; Owen and Friedenstein 1988). It was later modified to allow for the possibility of functional interchange between cells of different lineages, the best example being the conversion of marrow adipocytes into osteoblasts (Beresford et al.

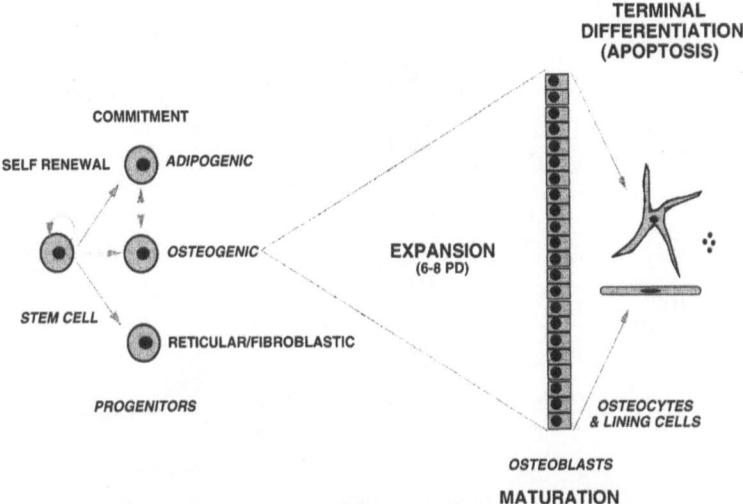

Fig. 1. A hypothetical view of the differentiation of marrow stromal cells of the osteogenic lineage

1992; Bennett et al. 1991) (Fig. 1). Whilst this has proved an attractive and highly influential model, definitive proof of the existence of stromal stem cells in adult bone marrow is still lacking. An alternative view, for which there is some experimental support, is that the size of the stem cell population becomes fixed at or shortly after birth (Waller et al. 1995) (Fig. 2). In this model the stem cells give rise to a single class of proliferative progenitors that, depending on the needs of the tissue, can differentiate into any of the functional marrow stromal cell types. It is envisaged that, even at a late stage during their development, these will retain considerable plasticity of phenotype, thus allowing for the marked and often rapid changes in the cellular composition of the marrow stroma that are known to occur following injury and in certain disease states (Bianco and Riminucci 1998).

The study of the more primitive cells of the marrow stromal system (CFU-F) has proved difficult because of their low incidence in normal human bone marrow (typically about 1 in 10^4 nucleated cells) and because, unlike their hematopoietic counterparts, they remain morphologically indistinguishable until a late stage during their development

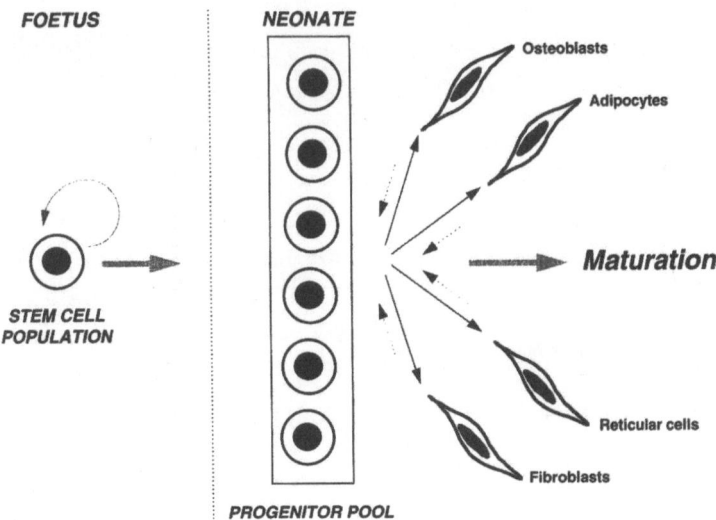

Fig. 2. An alternative view of marrow stromal cell differentiation. Further details are given in the text

(Beresford 1989). In cultures of marrow cell suspensions their presence can be shown by functional assay. CFU-F are highly adherent cells that, by a process of clonal expansion, give rise to colonies that vary in size and morphology. The number of colonies formed can therefore be used as an estimate of the number of CFU-F present in the original suspension (Beresford 1989; Friedenstein 1976, 1980). In a variation on this theme, the developmental potential of the CFU-F under the specific conditions of culture employed can be assessed by staining colonies for the expression of lineage specific markers, or by demonstrating the presence of cells with distinguishing morphological characteristics, most notably adipocytes (Owen et al. 1987; Locklin et al. 1995; Bennett et al. 1991; Malaval et al. 1994). In the case of cells with osteogenic potential, their presence can be shown by the formation of well organized, three-dimensional mineralized structures that histologically resemble woven bone, the "nodule-forming assay" (Maniatopoulos et al. 1988; Bellows and Aubin 1989; Aubin and Herbertson 1998). Few investigators, however, have applied the technically demanding, but ultimately most informative, procedure of cloning in vitro and transplantation in vivo, a technique devised originally by the late Alexander Friedenstein (Friedenstein 1980; Friedenstein et al. 1987; Kuznetsov et al. 1997).

An alternative to the use of functional assays to detect the presence of CFU-F and to assess the developmental potential of their immediate progeny is that of determining their degree of reactivity with antibodies recognizing lineage- and stage-specific antigens. The utility of this approach has been amply demonstrated in the field of hematopoiesis, in which the availability of a large number of antibodies recognizing "cluster of differentiation" (CD) antigens has greatly facilitated the establishment of lineage relationships in this complex system and enabled the isolation of specific subpopulations of cells for further investigation (Schlossman et al. 1997).

A number of antibodies have now been generated that recognize human cells of marrow stromal origin (Simmons and Torok Storb 1991; Stewart et al. 1996; Joyner et al. 1997; Bruder et al. 1997; Walsh et al. 1994). Of these, perhaps the most interesting is the monoclonal antibody STRO-1, which recognizes a cell surface, trypsin-resistant antigen expressed by a subset of human marrow stromal cells that includes essentially all assayable CFU-F (Simmons and Torok Storb 1991). In vitro, STRO-1+ CFU-F proliferate extensively and give rise to cells of multiple marrow stromal cell lineages, including osteoblasts (Simmons and Torok Storb 1991; Gronthos et al. 1994).

In this investigation, the expression of STRO-1 has been studied in cultures of adult human marrow stromal cells in relation to that of the bone/liver/kidney isozyme of alkaline phosphatase (AP), the expression of which is one of the earliest signs that a cell has initiated a program of osteogenic differentiation. The results show that in addition to STRO-1+ and AP+ cells, these cultures contain a subpopulation with the surface phenotype STRO-1+/AP+. It is proposed that these represent committed osteoprogenitors, that they differentiate from cells in the STRO-1+/AP− subpopulation and that they mature into osteoblast-like cells with the surface phenotype STRO-1−/AP+.

2.2 Methods

2.2.1 Cell Culture

Trabecular bone and bone marrow were obtained from ribs removed from patients with no prior evidence of bone disease that were undergoing cardiothoracic surgery. Normal human bone-derived cells (NHBC) (Gallagher et al. 1996) were grown from trabecular bone fragments and bone marrow stromal cells (BMSC) (Gronthos et al. 1998) from the mononuclear cell fraction of marrow recovered by density gradient centrifugation. Cells were cultured in Dulbecco's modification of Eagle's minimal essential medium (DMEM; high glucose formulation) supplemented with 10% (v/v) heat-inactivated fetal calf serum (FCS) from a selected batch and 100 μM L-ascorbic acid 2-phosphate (Alpha Labs, Farnham, Hants, UK; hereinafter referred to as standard medium) in the presence or absence of 10 nM dexamethasone (Dx; Sigma Chemical Co., Poole, Dorset, UK). NHBC were grown to confluence in primary culture (5–7 weeks), harvested and grown to confluence in secondary culture (2–4 weeks). BMSC were redistributed after 3 weeks in primary culture and grown to confluence in secondary culture (2–4 weeks). All experiments were performed using two cultures. Cells were harvested following sequential treatment with collagenase/trypsin-EDTA as described previously (Gallagher et al. 1996).

2.2.2 Labeling of Cells for STRO-1 and B4-78

A total of 10^5 cells were used for each of the controls and in excess of 5×10^6 cells for the dual-labeled sample. Cells were incubated in blocking buffer (1% BSA, 5% FCS, 10% normal human serum in Hank's buffered salt solution) for 30 min on ice and then reacted with combinations of the monoclonal antibodies STRO-1 and B4-78 (anti-B/L/K isozyme of AP; both obtained from the Developmental Studies Hybridoma Bank, University of Iowa, http://www.uiowa.edu/~dshbwww/), or isotype-matched control antibodies, on ice for 45 min. Antibody binding was detected using a combination of anti-mouse IgM FITC and anti-mouse IgG1 R-PE.

2.2.3 Flow Cytometry

Cells were analyzed using a Becton Dickenson FACStar Plus. The negative control sample was used to set quadrants on a dot plot of FL2 (R-PE) versus FL1 (FITC) such that 95%–100% of cells appeared in the lower left quadrant. Compensation levels were set using the STRO-1 and B4-78 controls. Sort regions were assigned using the dual-labeled sample to enable the recovery of cells which were of the phenotype STRO-1–/AP+, STRO-1+/AP+, STRO-1–/AP- or STRO-1+/AP– (Fig. 3). The sorted cells were reanalyzed to assess purity, counted and the percentage viability determined using trypan blue (routinely 90%).

Cells isolated by fluorescent activated cell sorting (FACS) were subcultured in standard medium into 6-well plates at a density of $10^4/cm^2$. After 7 days, the cells were harvested following a brief exposure to trypsin-EDTA and then re-labeled with the monoclonal antibodies STRO-1 and B4-78. Specific binding of the antibodies was then assessed by flow cytometry using 10^5 cells/test.

2.2.4 Analysis of mRNA Transcript Expression by RT-PCR

Sorted cells were recovered by centrifugation, snap frozen in liquid nitrogen and then stored at –80°C until use. The mRNA was prepared from 1.5×10^4 cells using a Pharmacia Quickprep Micro mRNA purifica-

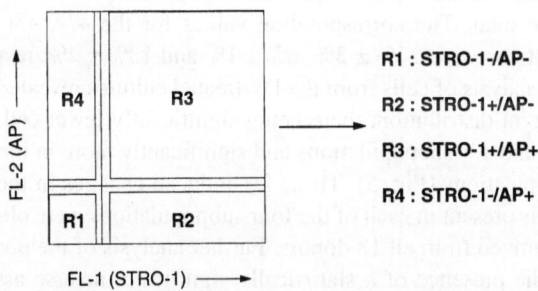

Fig. 3. The relationship between a cell's surface phenotype and the analysis/sort region (*R*) to which it is assigned when dual-labeled with the monoclonal antibodies STRO-1 and B4-78

tion kit. Half of the mRNA was then reverse transcribed in a reaction mix containing 10 U/ml Pd(N)$_6$, 0.8 mM dNTPs, 8 mM DTT, 30 U RNase inhibitor and 400 U M-MLV reverse transcriptase for 90 min at 37°C. The reaction was terminated by heating at 95°C for 20 min. Then, 1/20 of the cDNA was amplified using 2.5 U Taq polymerase in the presence of 0.5 mM dNTPs, 0.06–1 mM MgCl$_2$ and 1 mM of the relevant primer. The cycle number and conditions varied depending on the primer pairs used. Amplified products were separated by agarose gel electrophoresis, Southern blotted and then hybridized overnight at 65°C with end-labeled oligonucleotide probes that hybridize with sequences internal to those of the primers used in the PCR. Specifically bound probe was visualized by autoradiography. Images were captured using a CCD-camera and ImageDok software and then analyzed using the Phoretix 1-D analysis package. The transcript for GAPDH was used as an internal standard.

2.3 Results

Analysis of the dual-labeled cell populations showed the presence of four subpopulations defined on the basis of their cell surface expression of the STRO-1 antigen and/or AP: STRO-1–/AP–, STRO-1+/AP–, STRO-1+/AP+ and STRO-1–/AP+ (hereinafter referred to as –/–, +/–, +/+ and –/+ respectively; Fig. 4). Analysis of the pooled data from a large number of donors (n=18, 17 males and 1 female) revealed that, in control cultures, the –/– subpopulation comprised 54% ± 4% (mean ± SE) of the total. The corresponding values for the +/–, +/+ and –/+ subpopulations were 24% ± 3%, 6% ± 1% and 13% ± 2%, respectively (Fig. 5). Analysis of cells from the Dx-treated cultures revealed a markedly different distribution, there being significantly fewer cells present in the –/– and +/– subpopulations and significantly more in the +/+ and –/+ subpopulations (Fig. 5). These Dx-induced changes in the proportion of cells present in each of the four subpopulations were observed in cultures derived from all 18 donors. Further analysis of the pooled data revealed the presence of a statistically significant, inverse association between the proportion of cells present in the +/– (STRO-1+) and –/+ (AP+) subpopulations, which became much stronger in the presence of

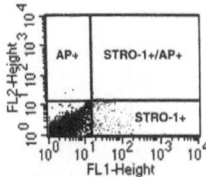

ISOTYPE-MATCHED CONTROL ANTIBODIES

Quad	Events	% Total	X Geo Mean	Y Geo Mean
UL	5	0.10	8.54	21.87
UR	102	2.04	73.79	40.15
LL	4705	94.10	3.10	2.18
LR	188	3.76	42.96	3.41

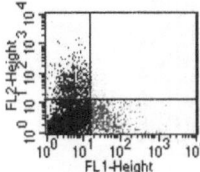

CONTROL

Quad	Events	% Total	X Geo Mean	Y Geo Mean
UL	636	12.72	4.92	44.95
UR	216	4.32	59.71	62.93
LL	3573	71.46	3.94	2.69
LR	575	11.50	40.78	3.38

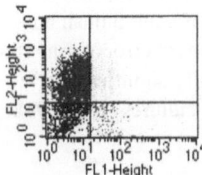

10 nM DEXAMETHASONE

Quad	Events	% Total	X Geo Mean	Y Geo Mean
UL	2625	52.50	6.46	74.97
UR	1487	29.74	36.76	103.00
LL	695	13.90	3.73	4.72
LR	193	3.86	47.29	4.30

Fig. 4. Identification of four subpopulations of human bone derived-cells using the monoclonal antibodies STRO-1 and B4-78 and two-color flow cytometry

Dx (Spearman's ρ and p-values of -0.692, <0.01 and -0.946, <0.001 for control and Dx-treated cultures respectively).

Cells from the $+/-$ and $-/+$ subpopulations were isolated by FACS. In each case the purity of the sorted populations was ~85% (Fig. 6). The recovered cells were recultured for 7 days and then reanalyzed for the expression of STRO-1 and/or AP. Post-culture, cells recovered in the $+/-$ fraction were found to be present in the $-/-$ and $+/-$ subpopulations in an approximately 1:1 ratio (Fig. 6). Some 53% of the cells recovered in the $-/+$ fraction retained the parental phenotype. Of the remainder, the majority were found to be present in the $-/-$ subpopulation with a small number (about 10% of total) present in the $+/+$ subpopulation (Fig. 6).

cDNA was prepared from cells recovered in each of the four sort regions and used as a template in PCR reactions with primers specific

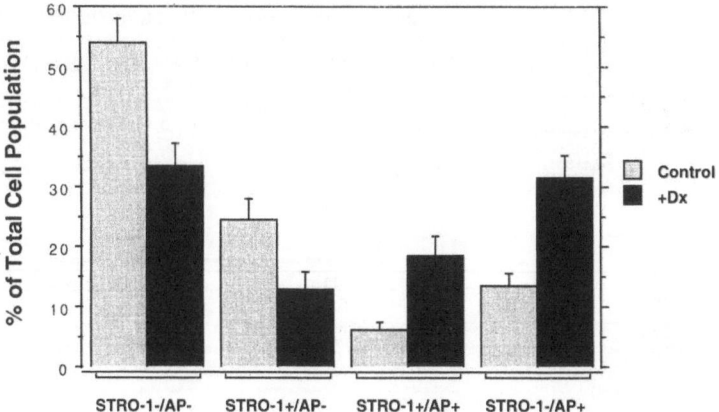

Fig. 5. The effect of treatment with dexamethasone (*Dx*) on the expression of the STRO-1 antigen and alkaline phosphatase in cultures of adult human bone marrow stromal cells. The data shown is the mean (± standard error) for n=18 donors. For each cell subpopulation there was a statistically significant difference between the control and dexamethasone-treated cultures (p <0.01). Further details are given in the text

for the parathyroid hormone receptor (PTHR) and the noncollagenous bone matrix protein bone sialoprotein (BSP). The products of the reactions were separated in agarose gels and then Southern blotted using oligonucleotide probes that hybridize with sequences internal to those of the PCR primers used. Transcripts for the PTHR were detected in cells from the –/+, +/+ and, at a low level, in the –/– subpopulations (Fig. 7). BSP transcripts were detected in the –/+, –/– and, at an extremely low level, in the +/+ subpopulations (Fig. 7).

2.4 Discussion

It is now widely accepted that in the postnatal mammal osteoblasts derive from multipotential precursors associated with the soft, fibrous tissue of the marrow stroma. In human bone marrow aspirates these cells are present within the STRO-1+, CFU-F population. Not all CFU-F are

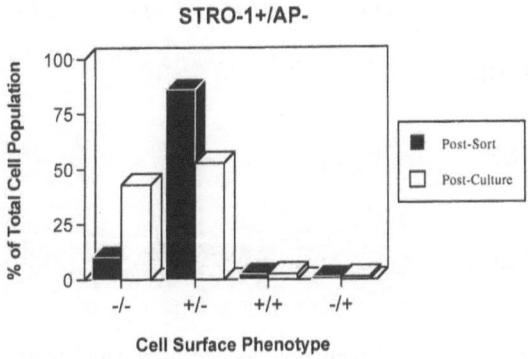

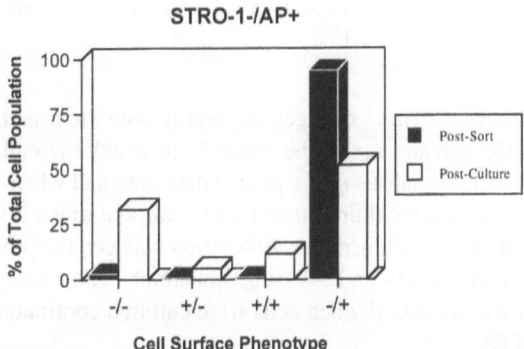

Fig. 6. Phenotypic stability of cells recovered in the STRO-1+/AP– or STRO-1–/AP+ fractions when recultured for 7 days in vitro

multipotential, however, and it is known that only a proportion are capable of undergoing osteogenic differentiation (Friedenstein 1980).

The results of this investigation show that STRO-1+ cells are present in cultures derived from human bone and marrow and that a proportion of these cells co-express AP, an early marker of osteogenic differentiation. When treated with Dx, a potent inducer of osteogenic differentiation in this cell culture system, the proportion of cells that express both STRO-1 and AP (+/+) is increased, as is the proportion that express AP alone (–/+). In contrast, the proportion of cells that express STRO-1

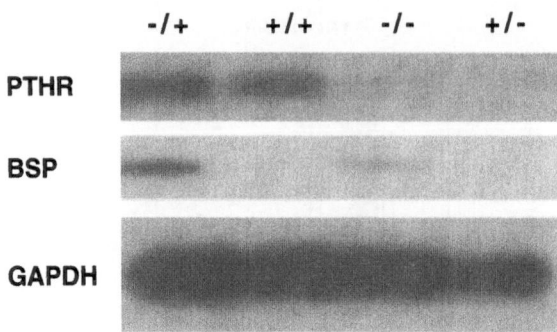

Fig. 7. The expression of transcripts for the parathyroid hormone receptor (*PTHR*) and bone sialoprotein (*BSP*) in the subpopulations of cells identified by dual labeling with the monoclonal antibodies STRO-1 and B4-78. Glyceraldehyde-3-phosphate dehydrogenase (GAPDH) was used as an internal standard

alone is decreased. These changes are highly consistent and are independent of the age and sex of the donor from which the cultures were derived. Statistical analysis of the pooled data obtained when cells from multiple donors were cultured under identical conditions revealed the presence of a significant, negative association between the proportion of cells present in the +/– and –/+ subpopulations. The strength of this association was increased when cells were cultured continuously in the presence of Dx.

Viable cells from the +/– and –/+ subpopulations could be isolated by FACS. When recultured, cells from both fractions proliferated. Analysis of the recultured cells after 7 days revealed a remarkable degree of phenotypic stability. Those recovered in the +/– fraction either retained the parental phenotype or ceased to express STRO-1. Similarly, the majority of cells recovered in the –/+ fraction retained the parental phenotype, with the remainder concentrated in the –/– fraction.

The expression of transcripts for two osteoblast-related genes, PTHR and BSP, was examined by RTPCR in cells recovered from all four sort regions. In the absence of Dx, transcripts for neither the PTHR nor BSP were detected in the +/– population. For cells recovered in the remaining fractions, the rank order of expression was –/+ ~ +/+ >> –/– and –/+ >> –/– >> +/+ for PTHR and BSP transcripts respectively.

Experimental Observations   **Hypothesis**

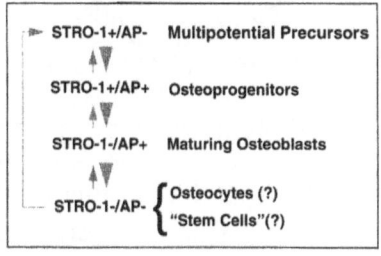

Fig. 8. The results of experiments in which the expression of transcripts for different genes has been analyzed by RT-PCR in the subpopulations of cells identified by dual labeling with the monoclonal antibodies STRO-1 and B4-78. Changes in the level of expression of transcripts for each gene are indicated by changes in the size of the characters used and in the type face (*bold vs plain text*). *BSP*, bone sialoprotein; *ER*, estrogen receptor; *FGFR*, fibroblast growth factor receptor; *OC*, osteocalcin; *PTHR*, parathyroid hormone receptor

Collectively, the results described are consistent with the possibility that the four subpopulations identified by dual labeling with the monoclonal antibodies STRO-1 and B4-78 contain cells of the osteoblast lineage at different stages of differentiation. On the basis of the results of these and other studies (summarized in Fig. 8), it is postulated that multipotential cells are present within the STRO-1+/AP– (+/–) subpopulation, that cells with the surface phenotype STRO-1+/AP+ (+/+) are committed osteoprogenitors and that the STRO-1+/AP+ subpopulation is comprised of maturing osteoblasts. It is proposed further that during the process of osteoblast differentiation cells progress from being STRO-1+/AP– to STRO-1+/AP+ and thence to STRO-1–/AP+.

The least well defined population of cells identified in this study have the surface phenotype STRO-1–/AP–. It is probable that of all of the subpopulations identified this is the most heterogeneous and that it includes very early (uncommitted) cells as well as late stage, AP-negative cells of the osteoblast lineage (osteocytes and/or lining cells). In

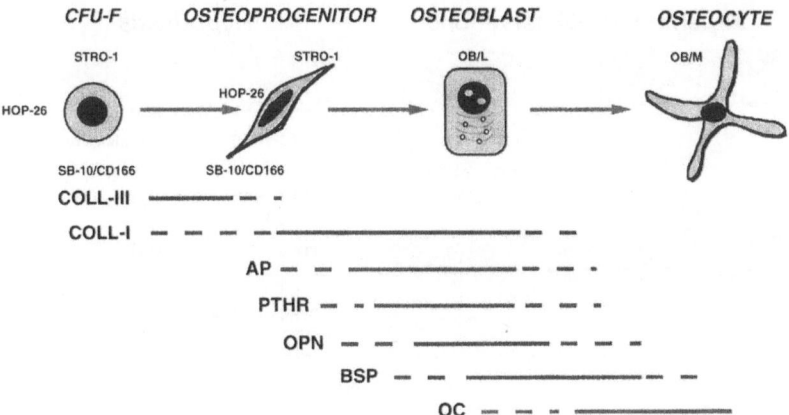

Fig. 9. Hypothetical and highly simplified schema for the differentiation of adult human cells of the osteoblast lineage. The differentiation process is depicted in relation to the expression of a subset of osteoblast-related genes and to the expression of selected cell surface and intracellular antigens. Further details in the text. For a more detailed and more generalized treatment of the subject the reader is referred to Aubin and Liu (1996)

future studies this possibility will be explored further using some of the additional stage- and lineage-specific antibodies that have been developed recently (Fig. 9) (Stewart et al. 1996; Joyner et al. 1997; Bruder et al. 1997; Walsh et al. 1994) as well as in vivo assays of the cells potential for proliferation and osteogenic differentiation (Krebsbach et al. 1997; Kuznetsov et al. 1997; Gundle et al. 1995).

Acknowledgements. This work was supported by grants from the Arthritis Research Campaign (WO525 & BO512) and the Bayer Corporation.

Dedication
This chapter is dedicated to the memory of Alexander J. Friedenstein.

References

Aubin JE , Liu F (1996) The osteoblast lineage. In: Bilezikian, JP, Raisz, LG, Rodan, GA (eds) Principles of bone biology. Academic, San Diego, pp 51–67

Aubin JE, Herbertson A (1998) Osteoblast lineage in experimental animals. In: Beresford JN, Owen ME (eds) Practical animal cell biology series: Marrow stromal cell culture. Cambridge University, Cambridge, UK, pp 88–110

Bellows CG, Aubin JE (1989) Determination of the number of osteoprogenitors present in isolated fetal rat calvaria cells in vitro. Dev Biol 133: 8–13

Bennett JH, Joyner CJ, Triffitt JT, Owen ME (1991) Adipocytic cells cultured from marrow have osteogenic potential. J Cell Sci 99: 131–139

Beresford JN (1989) Osteogenic stem cells and the stromal system of bone and marrow. Clin Orthop 240: 270–280

Beresford JN, Bennett JH, Devlin C, Leboy PS, Owen ME(1992) Evidence for an inverse relationship between the differentiation of adipocytic and osteogenic cells in rat marrow stromal cell cultures. J. Cell Sci: 102, 341–351

Bianco P, Riminucci M (1998) The bone marrow stroma in vivo: ontogeny, structure, cellular composition and changes in disease. In: Beresford JN, Owen ME (eds) Practical animal cell biology series: Marrow stromal cell culture. Cambridge University, Cambridge, UK, pp 10–25

Bruder SP, Horowitz MC, Mosca JD, Haynesworth SE (1997) Monoclonal antibodies reactive with human osteogenic cell surface antigens. Bone 21: 225–235

Cohen-Solal ME, Shih M-S, Lundy MW, Parfitt AM (1991) A new method for measuring cancellous bone erosion depth: Application to the cellular mechanisms of bone loss in postmenopausal osteoporosis. J Bone Miner Res 12: 1331–1337

Compston JE, Vedi S, Mellish RWE, Croucher, PE, O'Sullivan MM (1989) Reduced bone formation in non-steroid treated patients with rheumatoid arthritis. Ann Rheum Dis 48: 483–487

Friedenstein AJ (1976) Precursor cells of mechanocytes. Int Rev Cytol 47: 327–355

Friedenstein AJ(1980) Stromal mechanisms of bone marrow: Cloning in vitro and transplantation in vivo. In: Thienfelder S (ed) Immunobiology of bone marrow transplantation. Springer-Verlag, Berlin Heidelberg New York, pp 19–29

Friedenstein AJ, Chailalhyan RK, Gerasimov UV (1987) Bone marrow osteogenic stem cells: In vitro cultivation and transplantation in diffusion chambers. Cell Tiss Kinet 20: 263–272

Gallagher JA, Gundle R, Beresford JN (1996) Isolation and culture of bone-forming cells (osteoblasts) from human bone. In: Jones GE (ed) Methods in

molecular medicine: human cell culture protocols. Humana, Totowa, NJ, pp
233–262

Gronthos S, Graves SE., Ohta S, Simmons PJ (1994) The STRO-1(+) fraction
of adult human bone-marrow contains the osteogenic precursors. Blood 84:
4164–4173

Gronthos S, Graves SE., Ohta S, Simmons PJ (1998) Isolation, purification
and in vitro manipulation of human bone marrow stromal precursor cells.
In: Beresford JN, Owen ME (eds) Practical animal cell biology series: marrow stromal cell culture. Cambridge University, Cambridge, UK, pp 26–42

Gundle R., Joyner CJ, Triffitt JT (1995) Human bone tissue formation in diffusion chamber culture in vivo by bone-derived cells and marrow stromal fibroblastic cells. Bone 16: 597196601

Jilka RL., Weinstein RS, Takahashi K, Parfitt AM, and Manolagas, S.C (1996)
Linkage of decreased bone mass with impaired osteoblastogenesis in a murine model of accelerated senescence. J Clin Invest 97: 1732–1740

Joyner CJ., Bennett A, Triffit JT (1997) Identification and enrichment of human osteoprogenitor cells by using differentiation stage-specific monoclonal antibodies. Bone 21: 1–6

Krebsbach PH, Kuznetsov SA, Satomura K, Emmons RVB, Rowe DW, Robey
PG (1997) Bone formation in vivo: comparison of osteogenesis by transplanted mouse and human marrow stromal fibroblasts. Transplantation 63:
1059–1069

Kuznetsov SA, Krebsbach PH, Satomura K, Kerr J, Riminucci M, Benayahu
D, Robey PG (1997) Single-colony derived strains of human marrow stromal fibroblasts form bone after transplantation in vivo. J Bone Miner Res
12: 1335–1347

Locklin RM, Williamson MC, Beresford JN, Triffit JT, Owen ME (1995) In vitro effects of growth-factors and dexamethasone on rat marrow stromal
cells. Clin Orthop 313: 27–35

Malaval L, Modrowski D, Gupta AK, Aubin JE (1994) Cellular expression of
bone-related proteins during in-vitro osteogenesis in rat bone-marrow stromal cell-cultures. J Cell Physiol 158: 555–572

Maniatopoulos C, Sodek J, Melcher AH (1988) Bone formation in vitro by
stromal cells obtained from bone marrow of young adult rats. Cell Tissue
Res 254: 317–330

Owen ME (1985) Lineage of osteogenic cells and their relationship to the stromal system. In: Peck WA (ed) Bone and mineral research 3. Excerpta
Medica, Amsterdam, pp 1–25

Owen ME, Cave J, Joyner CJ (1987) Clonal analysis in vitro of osteogenic differentiation of marrow stromal CFU-F. J Cell Sci 87: 731–738

Owen ME and Friedenstein AJ (1988) Stromal stem cells : Marrow derived
osteogenic precursors. In: Cell and molecular biology of vertebrate hard tis-

sues. Ciba Foundation Symposium 136. John Wiley and Sons, Chichester, UK, pp 42–60

Parfitt AM (1990a) The three organizational levels of bone remodeling: Implications for the interpretation of biochemical markers and the mechanisms of bone loss. In: Christiansen C, Overgaard K (eds) Osteporosis 1990. Third international symposium on osteoporosis. Osteopress ApS, Copenhagen, pp 429–434

Parfitt AM (1990b) Bone-forming cells in clinical conditions. In: Hall BK (ed) Bone: a treatise, vol.1: The osteoblast and osteocyte. Telford, Caldwell, NJ, pp 351–429

Parfitt AM (1992) Implications of architecture for the pathogenesis and prevention of vertebral fracture. Bone 13: S41-S47

Schlossman, SF, Boumsell L, Kishimoto T, Knapp W, Mason D, McMichael A, Shaw S, Springer TA, Tedder T. Todd R, Waldman H, Zola H (1997) CD antigens 1996: updated nomenclature for clusters of differentiation on human cells. Bull World Health Org 75: 385–387

Simmons PJ, Torok Storb B (1991) Identification of stromal cell precursors in human bone marrow by a novel monoclonal antibody, STRO-1. Blood 78: 55–62

Stewart K, Gronthos S, Simmons PJ, Beresford JN (1996) H8G – a monoclonal-antibody that subtypes the STRO-1+ CFU-F population of human bone and marrow. J Bone Miner Res 11: p. 19

Waller EK, Huang S, Terstappen L (1995) Changes in the growth-properties of CD34(+), CD38(-) bone-marrow progenitors during human fetal development. Blood 86: 710–718

Walsh S, Dodds RA, James IE, Gowen M (1994) Monoclonal antibodies with selective reactivity against osteoblasts and osteocytes in human bone. J Bone Miner Res 9: 1687–1696

The Identification and Role of the Cell of the Intestinal Epithelium

3 Apoptosis in Bone Cells

B.F. Boyce, D.E. Hughes, K.R. Wright, L. Xing, and A. Dai

3.1 Introduction

Apoptosis is an important regulatory process that, in combination with cell division, determines not only cell numbers, but also the shape and volume of individual organs (Kerr et al. 1972; Wyllie et al. 1980). It is critical during embryonic development for the elimination of superfluous cells, such as the connective tissue between developing fingers (a bone morphogenetic protein (BMP)-dependent process; Zou and Niswander 1996; Yokouchi et al. 1996), and for ensuring that folding and rotation of primitive tissues occur at precise times. High mitotic and low apoptotic rates ensure the rapid growth of highly malignant tumors (Mooney et al. 1995), while induction of massive apoptosis is a major mechanism of action of most chemotherapeutic agents (Lowe et al. 1993) and irradiation therapy (Lee and Bernstein 1993) to reduce tumor volume. Large-scale apoptosis of breast cells follows the cessation of lactation and leads to the associated reduction in breast size, and shrinkage of prostatic tissue after orchidectomy occurs by a similar mechanism (Reed 1994). Because bone is a hormonally sensitive tissue, it is possible that apoptosis of bone cells could be a major determinant of bone shape and volume and, as with the breast and prostate, the amount of tissue present is likely to be regulated by the activity of hormones and growth factors.

Although the term apoptosis, was coined in 1972 (Kerr et al. 1972), research on its regulation at the cellular and molecular level by scientists did not begin on a large scale until the late 1980s. Indeed, apoptosis of bone cells was not recognized until 1993, when it was reported that osteoclasts undergo apoptosis (Boyce et al. 1993; Fuller et al. 1993a, b). Since then, further studies have revealed that apoptosis is the fate of all cell types in bone and that it appears to be regulated by many of the factors that stimulate or inhibit bone turnover.

This chapter will briefly review the morphological features and regulation of apoptosis and the methods for detecting it. It will also review the studies of apoptosis in bone cells published to date, with the emphasis being on the possible role of this process as a determinant of bone shape, volume and strength.

3.2 Morphology and Detection

Apoptosis was defined morphologically by means of light and electron microscopy (Kerr et al. 1972; Wyllie et al. 1980) on the basis of characteristic nuclear and cytoplasmic changes (Fig. 1) which differ from those of cells undergoing ischemic or toxic necrosis. In contrast to necrotic cells, which take in fluid and swell but maintain their nuclear chromatin before bursting and eliciting an inflammatory reaction, apoptotic cells lose fluid, contract and detach from neighboring cells or matrix, while their nuclei condense and later disintegrate. Apoptotic cells break up into small apoptotic bodies which are phagocytosed by their neighbors (Fig. 2) or by phagocytes and do not elicit an inflammatory response. It is often difficult to recognize these apoptotic bodies in histologic sections, particularly if they do not contain nuclear fragments,

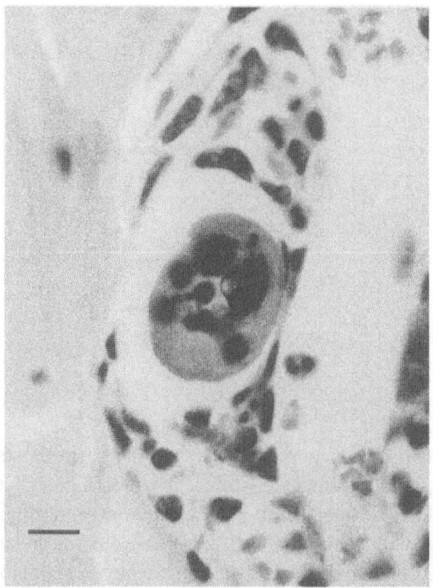

Fig. 1. Apoptotic osteoclast. The osteoclast in the center of this micrograph illustrates the morphological features of apoptosis, i.e., nuclear condensation and fragmentation, cytoplasmic contraction, and detachment of the cell from the bone and surrounding cells. Hematoxylin and eosin; scale bar, 8.5 μm

because they typically are small and blend in with the cytoplasm of surrounding or engulfing cells.

Apoptosis is a relatively uncommon event and, even in tissues such as the epithelial lining of the gastrointestinal tract where the cell turnover rate is high, as few as three in 1000 cells can be identified as being apoptotic (Lee 1993). Here, they can be identified relatively easily because there is a sufficiently large number of cells present on the mucosal surface to permit detection of the small number of dying cells. As will be seen later, detection of apoptosis in bone remodeling units in bone sections is more challenging because of the relatively small numbers of cells present.

Apoptotic cells also lose attachment not only to matrix, but also to culture dishes. This property has been used to collect such cells and quantify apoptosis in vitro (Arends et al. 1990; Hughes et al. 1996) and

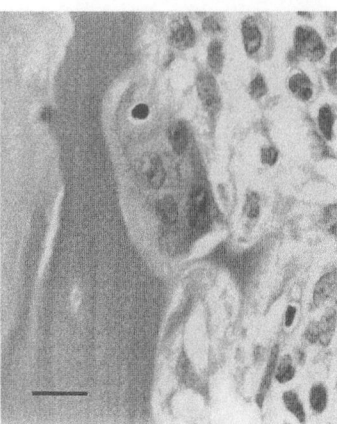

Fig. 2. Phagocytosis of an apoptotic osteocyte. The densely stained, contracted nucleus of an apoptotic osteocyte is present within a clear lysosomal vacuole of the multinucleated osteoclast in a Howship's resorption lacuna. The nuclei of the osteoclast with their finely dispersed chromatin and nucleoli contrast sharply with the condensed nucleus of the osteocyte. Hematoxylin and eosin; scale bar, 8.5 µm

has been recognized for many years by cell biologists as a feature of the behavior of transformed cells, which round up and detach from culture plates after reaching confluence. This intrinsic apoptosis in malignant cells must be considered in the analysis of any observed effects when transformed cells are being used as surrogates for normal cells to study the regulation of apoptosis in bone.

Fragmentation of nuclear chromatin is carried out by endogenous endonucleases which are activated in the process (Kerr et al. 1972; Wyllie 1980) and split genomic DNA into nucleosomal fragments of varying sizes using a Ca^{2+}/Mg^{2+}-dependent mechanism (Arends et al. 1990). These fragments of DNA can be detected on gel electrophoresis where they form characteristic ladders (Wyllie 1980) which reflect the variable electrophoretic mobility of the fragments but do not permit quantitation of apoptosis of particular cell types in mixed cell populations. Other techniques have been developed more recently to identify and quantitate apoptosis in cells in tissue culture and in histological sections. Hoechst (Arndt-Jovin and Jovin 1977), acridine orange

(Arends et al. 1990) and propidium iodide are used routinely to identify apoptotic cells in vitro because the condensed nuclear chromatin of these cells has a stronger fluorescent signal than the loosely dispersed chromatin of viable cells. Disappointingly, in our experience a positive signal is not seen with these techniques in a significant proportion of cells in culture which are clearly apoptotic, perhaps because the chromatin in some cells may become so condensed that chemical binding to it becomes impossible.

More recently, in situ nick end-labeling techniques have been developed such as TUNEL (TdT-mediated dUTP-biotin nick end-labeling) (Gavrieli et al. 1992) in which labeled nucleotides are attached to the free ends of DNA fragments using terminal transferase and are detected under light or fluorescence microscopy. These techniques are now widely used in many laboratories to detect apoptosis in cells in culture or in tissue sections (Fig. 3) and apoptosis may be quantified automatically using flow cytometry (Gold et al. 1993). These methods also underestimate the number of apoptotic cells and in our experience detect only about 50% of apoptotic osteoclasts in tissue sections (Wright et al. 1995), perhaps for the same reasons that the fluorescent dyes underestimate apoptosis.

The earliest stages of apoptosis can be identified using labeled annexin V. During the early stages of apoptosis, before nuclear or cytoplasmic changes can be seen morphologically, phosphatidylserine is moved from the inner to the outer surface of the plasma membrane where it appears to be involved in signaling the demise of the cell to neighboring cells or phagocytes which subsequently engulf it (Fadok et al. 1992; Martin et al. 1995). Annexin V binds to phosphatidylserine and this property has been exploited to produce a new assay for the detection of apoptotic cells (Vermes et al. 1995) that is available as a commercial kit.

3.3 Regulation of Apoptosis

Understanding of the regulation of apoptosis has advanced at a rapid pace over the last 5 years. Numerous death and anti-death genes have been identified and the signaling pathways that are activated or suppressed in response to survival or death signals are being studied intensively. Most cell types appear to possess a death-inducing mechanism

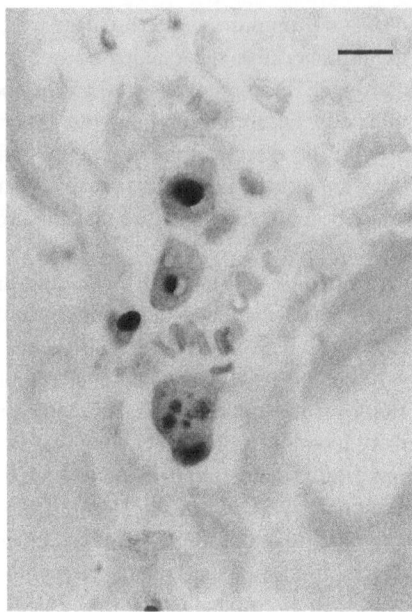

Fig. 3. TdT-mediated dUTP-biotin nick end-labeling (TUNEL) staining of apoptotic osteoclast nuclei. A strong, positive signal is present in some, but not all, of the nuclei of the three apoptotic osteoclasts which have retracted from the bone surface and are present in clear spaces in the center of this micrograph. In particular, only one of the three nuclei in the middle osteoclast has a positive signal. A strong signal is also present in a mononuclear cell of undetermined origin to the bottom and left of this cell. Methylene green and eosin counterstain; scale bar, 7 μm

and require signals from their neighbors or matrix to prevent them from triggering it by default. Major regulators include the Bcl-2 family of proteins, which includes pro- and anti-apoptotic members (Reed 1994), and interleukin-1 converting enzyme (ICE)-like cysteine proteases (Patel et al. 1996), now called caspases (reviewed by Villa et al. 1998), which are effectors of apoptosis. A detailed review of current knowledge of the molecular regulation of apoptosis is beyond the scope of this chapter; the subject is reviewed in Hale et al. (1996), Hoffman and Liebermann (1994) and White (1996). An examination of the list of

Table 1. Inducers of apoptosis

Physiological mechanisms	
Hormones	Glucocorticoids, sex hormones
Hormone/growth factor withdrawal	Bcl-2-mediated
Growth factors/cytokines	TGF-β, TNF, (Fas ligand)
Chemotherapy	Most cytotoxic drugs
Loss of matrix attachment	

Cell damage-related	
Free radicals	Oxygen and nitrogen, hydrogen peroxide
Cytotoxic T cells	Fas/TNF receptor and granzyme b
Irradiation	p53-mediated
Chemotherapeutic agents	
Oncogenes	Myc, Max, Mad, Fos
Tumor suppressors	p53

TGF, transforming growth factor; TNF, tumor necrosis factor.

Table 2. Inhibitors of apoptosis

Trophic hormones	
Sex steroids	
Growth factors	CSFs, IGF-I, NGF
Cytokines	Most interleukins, TNF
Bcl-2, Bcl-x$_L$	
Extracellular matrix	
Cysteine protease inhibitors	Caspases
Ca^{2+} channel blockers	
Calpain inhibitors	
Viral proteins	Adenovirus E1B 19 kDa; baculovirus p35 and 1AP; cowpoxvirus CrmA (serpin); Epstein Barr virus BHRF1 and LMP-1

CSF, colony-stimulating factor; IGF, insulin-like growth factor; NGF, nerve growth factor; TNF, tumor necrosis factor.

promoters and inhibitors of apoptosis in Tables 1 and 2, respectively, reveals that many of these agents have been identified already as stimulators or inhibitors of bone cell function. The effects of some of these agents on individual types of bone cells will be reviewed in subsequent sections.

3.4 Apoptosis in Bone Cells

3.4.1 Chondrocytes

Long bones form by endochondral ossification in which proliferating chondroblasts give rise to hypertrophic chondrocytes in growth plates situated near the ends of the bones. Recent studies have shown that hypertrophic chondrocytes undergo apoptosis before they and the surrounding calcified matrix are removed by osteoclasts (Lewinson and Silbermann 1992; Roach et al. 1995), and studies in knockout mice have revealed that parathyroid hormone-related protein (PTHrP) and possibly tartrate-resistant acid phosphatase (TRAP) play critical roles in this highly regulated process. PTHrP knockout mice have thinner than normal growth plates with reduced numbers of hypertrophic chondrocytes (Karaplis et al. 1994; Lee et al. 1996), while transgenic mice overexpressing PTHrP have thick growth plates and increased numbers of hypertrophic chondrocytes (Weir et al. 1996). PTHrP appears to control the life span of hypertrophic chondrocytes by preventing their premature apoptosis in a signaling mechanism which involves Indian hedgehog protein upstream of PTHrP (Lanske et al. 1996; Vortkamp et al. 1996) and Bcl-2 downstream (Amling et al. 1997). TRAP knockout mice have a phenotype with similarities to those seen in mice overexpressing PTHrP, i.e., they have thickened, distorted growth plates (Hayman et al. 1996). TRAP may therefore have an important function in regulating the activity of hypertrophic chondrocytes by limiting their life spans and counteracting the effects of PTHrP. Thus, the volume and shape of growth plates in developing long bones appears to be determined by the activity of factors which promote or prevent apoptosis of chondrocytes.

3.4.2 Osteoclasts

Osteoclasts are required during endochondral ossification for the removal not only of calcified cartilage at the growth plate, but also of much of the bone formed in the primary and secondary spongiosae by osteoblasts. This latter activity ensures the maintenance of a bone marrow cavity within long bones. Failure of osteoclast formation or function leads to the development of osteopetrosis, a condition in which bone volume is increased within the bone marrow cavity due to the accumulation of unresorbed osteocartilaginous matrix (Fig. 4) (Popoff and Marks 1995). Many examples of naturally occurring forms of osteopetrosis have been identified, and osteopetrosis has been the unexpected phenotype in c-*src* (Soriano et al. 1991), c-*fos* (Grigoriadis et al. 1994; Wang et al. 1992), PU.1 (Tondravi et al. 1997), and NF-κB (Franzoso et al. 1997; Iotsova et al. 1997) knockout mice as well as in transgenic (TRAP/Tag) mice expressing SV40 large T antigen under the control of the TRAP promoter (Fig. 4) (Boyce et al. 1995). Osteopetrosis occurs in *src* knockout mice because their osteoclasts do not form ruffled borders (Boyce et al. 1992) and in c-*fos*, PU.1 and NF-κB knockout mice because osteoclasts fail to form. Osteopetrosis could also occur if osteoclast activity was impaired as a result of increased osteoclast apoptosis. Osteoclast apoptosis was increased in an osteopetrotic line of TRAP/Tag transgenic mice expressing T antigen, but it was also increased in a nonosteopetrotic line of these mice, and we attributed the development of the osteopetrosis to impaired ruffled border formation associated with high levels of T antigen expression, rather than to increased osteoclast apoptosis (Boyce et al. 1995).

In studies done in collaboration with Drs. Schwartzberg and Varmus, at the National Institutes of Health, to further explore the role of *src* in osteoclasts, we have found that increased osteoclast apoptosis is associated with an increase in the severity of the osteopetrosis in *src* knockout mice expressing mutated or truncated *src* transgenes targeted to the osteoclast using the TRAP promoter. In these studies, we observed that expression of a wild-type *src* transgene rescues the osteopetrosis in *src* mutant mice and that a mutated *src* transgene which lacks kinase activity had a similar effect in some, but not all, of the mice expressing the transgene (Schwartzberg et al. 1997). Some of these transgenic mice actually developed more severe osteopetrosis than the *src* knockout

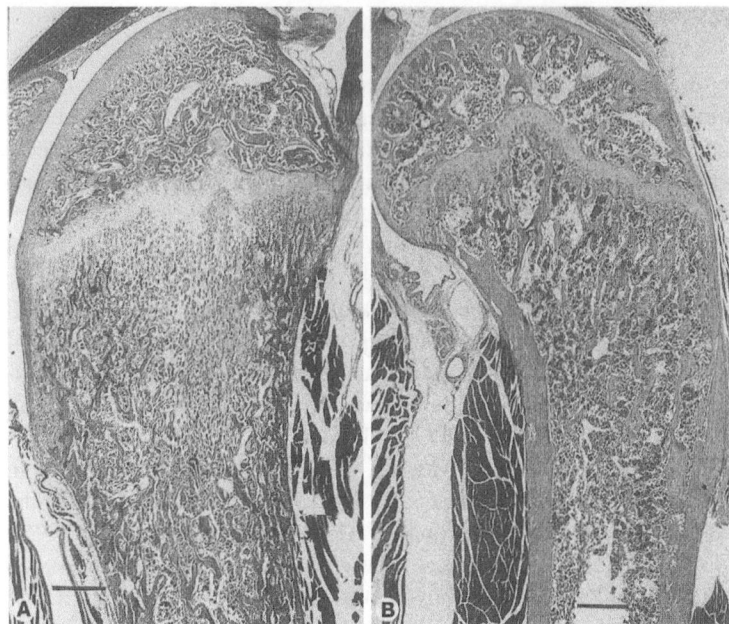

Fig. 4A, B. Osteopetrosis in the femur of a mouse. The bone marrow cavity of
the lower end of the femur of this osteopetrotic tartrate-resistant acid phosphatase (TRAP)/T antigen (Tag) transgenic mouse is largely filled with unremo-
deled osteocartilaginous matrix (**A**) and contrasts with the normal appearance
of a femur from a wild-type control (**B**). The osteopetrosis in these transgenic
mice is associated with the presence of morphologically transformed osteoclasts which have reduced ruffled border formation and increased numbers of
apoptotic osteoclasts. Hematoxylin and eosin; scale bar, 360 μm

mice and this was associated with increased osteoclast apoptosis (Xing
et al. 1997). Furthermore, in parallel studies we found that severe
osteopetrosis developed in all *src* knockout mice expressing a truncated
src transgene which lacks the kinase domain but encodes the first 251
amino acids, including the SH3 and SH2 domains which typically bind
interacting molecules (Xing et al. 1996). Osteoclast apoptosis was
greatly increased in these mice and also in some wild-type and *src*
heterozygote mice expressing this transgene which developed mild or

moderate osteopetrosis. Our hypothesis to explain these findings is that both of these *src* transgenes are acting in a dominant-negative manner to interfere with the function of interacting molecules that maintain survival of osteoclasts in *src* knockout mice. We do not yet know what these molecules are, but they could be other members of the *src* family or their substrates. Whatever the mechanism, these studies clearly demonstrate that these forms of osteopetrosis are associated with increased osteoclast apoptosis and suggest that reduction in the life span of osteoclasts can lead to increased bone mass in growing mice. It remains to be determined whether any forms of human osteopetrosis are due to increased osteoclast apoptosis.

Regulation of osteoclast life span could also be an important determinant of bone mass in the adult skeleton, for example before and after the menopause, if estrogen affects osteoclast survival. Estrogen withdrawal after the menopause or following ovariectomy is associated with increased generation and activity of osteoclasts which appear to be mediated by the cytokines interleukin (IL)-6, IL-1 and tumor necrosis factor (TNF) (Horowitz 1993; Pacifici 1996). Recent studies have shown that IL-1 enhances osteoclast survival (Jimi et al. 1995) and we have found that IL-1, IL-6 and TNF all prevent osteoclast apoptosis (Hughes et al. 1994a). Thus, bone loss after the menopause may result from a combination of the effects of increased osteoclast generation, activity and survival. Enhanced survival could permit osteoclasts to remain on bone surfaces longer than before the menopause and thus lead to the excavation of deeper than normal resorption lacunae and ultimately to the full-thickness resorption through trabeculae which is characteristic of post-menopausal osteoporosis.

If estrogen withdrawal can result in prolongation of osteoclast life span through the effects of these cytokines, it is also possible that estrogen could limit the life span of osteoclasts before the menopause and when given as hormone replacement therapy by promoting osteoclast apoptosis. We have found that estrogen promotes osteoclast apoptosis both in vitro and in vivo when given to mice following ovariectomy and that this effect appears to be mediated by transforming growth factor-β (TGF-β) (Hughes et al. 1996). Tamoxifen, which has estrogen agonistic effects on bone cells, also promotes osteoclast apoptosis (Arnett et al. 1996; Hughes et al. 1996; Lutton et al. 1996), an effect like that of estrogen which was prevented by TGF-β antibody (Hughes et al.

1996). Orchidectomy and hypogonadism are associated with osteoporosis and bone loss, which appears to be mediated by cytokine-stimulated bone resorption (Bellido et al. 1995). Testosterone also promotes osteoclast apoptosis in vitro and in vivo (Hughes et al. 1995a). Thus, a significant component of the protective effect of sex steroids in maintaining bone mass in both females and males may be limitation of osteoclast life span on bone surfaces to prevent them from remaining on trabeculae long enough to cause perforating resorption and its associated reduction in bone mass and strength.

The other major antiresorptive agents used to treat or prevent post-menopausal osteoporosis are bisphosphonates and calcitonin. While the precise mechanism of action of bisphosphonates to inhibit osteoclastic bone resorption remain unclear, our recent findings that bisphosphonates promote osteoclast apoptosis in a dose-dependent fashion (Hughes et al. 1995b) suggest that this may be a major mechanism of action of these highly effective anti-osteoclast agents. Induction of osteoclast apoptosis by them does not appear to be mediated by TGF-β (Selander and Boyce, unpublished observations), but may involve inhibition of the prenylation of GTP-binding proteins, such as Ras (Luckman et al. 1998), by isoprenoids which are products of the mevalonate pathway of cholesterol biosynthesis (Hughes et al. 1997).

Previous studies have shown that calcitonin has a transient, inhibitory effect on osteoclast mobility and a longer-term effect to inhibit bone resorption and prevent post-menopausal bone loss. Recent studies (Hughes et al. 1994b; Selander et al. 1996) indicate that calcitonin causes loss of attachment of osteoclasts to culture wells or dentine without inducing them to undergo apoptosis. Thus, although the precise mechanism of action of calcitonin remains unknown, its inhibitory effect on osteoclast function does not lead inevitably to osteoclast apoptosis.

Recent studies have indicated that other inhibitors of bone resorption also induce osteoclast apoptosis. For example, Kameda et al. (1995) found that protein kinase C signaling is involved in neonatal rabbit osteoclast apoptosis and that vitamin K2, but not vitamin K1 (Kameda et al. 1996), induces osteoclast apoptosis. Lutton et al. (1996) have found that cyclosporine A, tamoxifen, corticosterone and dexamethasone all induce osteoclast apoptosis, with cell shrinkage becoming detectable under time-lapsed video photography within 25 min and marked morphological changes in osteoclasts within 2–4 h of treatment.

More recently, inhibition of NF-κB signaling in has been reported to be associated not only with inhibition of bone resorption by mature rabbit osteoclasts, but also with induction of osteoclast apoptosis (Ozaki et al. 1997). NF-κB is a family of transcription factors which regulate the expression of cytokines, such as IL-1, IL-6, and TNF (Siebenlist et al. 1994), and has been shown previously to have anti-apoptotic effects in some cells types (Beg and Baltimore 1996). IL-1 activates NF-κB signaling in mouse osteoclasts (Jimi et al. 1996), suggesting that NF-κB may be involved in the anti-apoptotic effect of this cytokine. We found no evidence of increased osteoclast apoptosis in NF-κB knockout mice which develop severe osteopetrosis (Franzoso et al. 1997). In these mice which lack expression of the p50 and p52 subunits of NF-κB, osteoclasts and their mononuclear TRAP-positive precursors fail to form. We do not yet know at which stage in osteoclast development the block in differentiation occurs or the mechanism involved, but it is possible that apoptosis of mononuclear TRAP-negative osteoclast precursors is the basis of the defect in these mice.

3.4.3 Osteoblasts

Osteoblasts are recruited to the primary and the secondary spongiosae in growth plates and to sites of previous resorption in bone remodeling units. Recruitment of osteoblasts to growth plates has not been studied in detail, but it has long been recognized that more osteoblasts appear to be recruited to bone remodeling units than ultimately end up as osteocytes in matrix or as lining cells on the bone surface (Parfitt 1994). It is likely that these osteoblasts undergo apoptosis, but to date there are still no published reports of osteoblasts undergoing apoptosis during normal bone modeling or remodeling.

We have attempted to identify apoptotic osteoblasts in bone remodeling units using morphological analysis and TUNEL staining in a variety of settings, including the mouse calvaria following treatment with IL-1 and parathyroid hormone (PTH), and in bone remodeling units in bone biopsy specimens from patients with primary and secondary hyperparathyroidism and Paget's disease, but so far have not observed osteoblasts undergoing apoptosis in these situations in any systematic fashion. The reason for our failure to observe osteoblast apoptosis in bone remodeling units may be technical. In regenerating tissue, such as the intestinal mucosa, only one to five in 1000 enterocytes appear apoptotic (Lee 1993). These apoptotic cells can be recognized readily because many thousands of enterocytes can be seen in a typical section of bowel mucosa. In contrast, the number of osteoblasts on the surface of bone remodeling units in a bone section may be only 200–300. Thus, if two or three in 1000 osteoblasts are apoptotic at any time during bone formation, the chances of observing apoptosis in a single section cut through a bone remodeling unit are extremely small, particularly if, as in other small cells, apoptosis in osteoblasts lasts for less than 2 h. Nevertheless, occasional TUNEL-positive osteoblasts have been reported recently at sites of bone formation in the secondary spongiosa of mouse femoral metaphyses (Jilka et al. 1998). The significance of osteoblast apoptosis in bone remodeling units, if it does occur, is that it could be increased with increasing age and result in fewer osteoblasts surviving to fill in resorption lacunae with an associated reduction in mean wall width and therefore in bone strength.

There are now several reports of osteoblasts undergoing apoptosis in vitro. For example, cultured MC3T3-E1 osteoblastic cells undergo apoptosis in response to TNF-α, an effect which involves ceramide-induced translocation of NF-κB from the cytoplasm to the nucleus (Kitajima et al. 1996). Nitric oxide also induces apoptosis of MC3T3-E1 cells in vitro, an effect which is enhanced by addition of TNF-α and interferon-γ (Damoulis and Hauschka 1997). TNF-α-induced osteoblast apoptosis is prevented by TGF-β and cytokines that utilize the gp130 signal transducer, such as IL-6 and leukemia inhibitory factor (Jilka et al. 1998). Thus, the inhibition of bone formation that typically accompanies increased bone resorption at sites of inflammation in bone could be due in part at least to promotion of osteoblast apoptosis by these agents and involve the ceramide-NF-κB signaling pathway.

3.4.4 Osteocytes

Osteocytes are long-lived cells which reside within bone matrix from the time of their incorporation into newly formed osteoid until they are removed from bone during osteoclastic resorption or until they die in situ (Dunstan et al. 1990). Although the fate of osteocytes has been the subject of speculation for many years, it now appears that at sites of bone resorption some, if not all, of these cells undergo apoptosis and are phagocytosed by osteoclasts (Bronckers et al. 1996; Elmardi et al. 1990; Hughes et al. 1996).

The precise function of osteocytes within bone remains unclear and is the subject of speculation. For example, they may function as strain transducers within bone and provide signals to osteoclasts from sites that are no longer under strain because of changes in strain patterns (Lanyon 1993). Such areas of bone could then be removed and replaced by new bone laid down along new lines of strain. They may also signal to osteoclasts sites of bone which have undergone microdamage and require replacement. Their death within bone could either be a signal to osteoclasts to remove areas of effete bone or else result in failure of communication between these two cell types and in the accumulation of areas of dead, weakened bone.

Osteocyte disappearance from lacunae has been reported to be increased in trabecular bone in femoral heads of elderly subjects with hip fractures and to increase with age at this site, but not in vertebral cancellous bone (Dunstan et al. 1990, 1993). The cause of the disappearance of osteocytes from their lacunae in these studies was not clear, but it is likely that the cells died by apoptosis. These findings of empty osteocyte lacunae have been confirmed recently in iliac bone biopsies by Mullender at al. (1996), but they found that there was no increase in the percentage of empty lacunae with age or in osteoporotic subjects at this site compared with age-matched controls (mean values were 17%–22%). Although this latter finding does not support a role for osteocyte death leading to defective mechano-sensing in osteoporosis, it does not refute it either because the study was done in subjects after they had developed osteoporotic fractures. Increased osteocyte death may have been present some years earlier when the bone loss that led to osteoporosis was taking place and the dead bone removed permanently. These authors did report that osteocyte numbers were increased per unit bone area in osteoporotic vs control subjects and suggested that osteoblasts in osteoporotic subjects lay down less bone matrix per cell than osteoblasts in normal subjects.

Recent studies have suggested that osteocytes undergo apoptosis in response to treatment with gonadotrophin releasing hormone analogues (Tomkinson et al. 1997a), an effect which has been attributed to the associated reduction in circulating estrogen levels in treated patients. In this study, nick translation or fluorescent dyes were the techniques used predominantly to identify osteocytes considered to be undergoing apoptosis in frozen sections of bone. Cells with a positive signal using these techniques appeared to have a random distribution in bone among viable cells. Thus, their distribution did not appear to correspond to that of a group of osteocytes which would likely be removed subsequently en-bloc during excavation of Howship's resorption lacunae. In preliminary studies, these investigators have found positive signals with these techniques in osteocytes in metaphyseal bone sections of rats following ovariectomy, but not in osteocytes of rats given estrogen after ovariectomy (Tomkinson et al. 1997b). It is not yet clear what percentage of such cells identified with these techniques is actually apoptotic as determined by the use of standard morphologic techniques. Further studies will be required to determine whether estrogen deficiency following ovariectomy or the menopause leads to osteocyte death and whether this cell demise is a signal for osteoclastic resorption or a response of osteocytes to imminent osteoclastic resorption.

3.5 Summary

The observation that bone cells die by apoptosis has opened up a new field of investigation into the regulation of bone modeling and turnover. It should lead to the development of new therapeutic agents designed specifically to promote or inhibit apoptosis in particular cell types with the aim of preventing bone loss or promoting bone formation. For example, identification of the targets in osteoclasts which mediate the induction of osteoclast apoptosis by estrogen, tamoxifen and the newer selective estrogen receptor modulators, such as raloxifene and droloxifene, could result in the development of drugs which use this mechanism specifically to inhibit bone loss after the menopause without inducing unwanted stimulatory effects on other tissues, such as the breast and endometrium. Inhibition of premature osteoblast death by apoptosis could result in the production of more bone matrix per osteoblast during bone formation and thus to thicker, stronger trabeculae, while inhibition of osteocyte apoptosis could result in prolongation of the integrity of bone and thus prevention of fractures.

*Acknowledgments.*We thank Tonya Keys for typing the manuscript. Part of the work described in these studies was supported by grants from the National Institutes of Health of the United States (AR 43510, AR 41336 and DK 45229).

References

Amling M, Neff L, Tanaka S, Inoue D, Kuida K, Weir E, Philbrick WM, Broadus AE, Baron R (1997) Bcl-2 lies downstream of parathyroid hormone-related peptide in a signaling pathway that regulates chondrocyte maturation during skeletal development. J Cell Biol 136:205–213

Arends MJ, Morris RJ, Wyllie AH (1990) Apoptosis: The role of the endonuclease. Am J Path 136:593–608

Arndt-Jovin DJ, Jovin TM (1977) Analysis and sorting of living cells according to deoxyribonucleic acid content. J Histochem Cytochem 25:585–589

Arnett TR, Lindsay R, Kilb JM, Moonga BS, Spowage M, Dempster D (1996) Selective toxic effects of tamoxifen on osteoclasts: comparison with the effects of estrogen. J Endocrinol 149:503–508

Beg AA, Baltimore D (1996) An essential role for NF-kB in preventing TNF-α-induced cell death. Science 274:782–784

Bellido T, Jilka RL, Boyce BF, Girasole G, Broxmeyer H, Dalrymple SA, Murray R, Manolagas SC (1995) Regulation of interleukin-6, osteoclastogenesis, and bone mass by androgens. J Clin Invest 95:2886–2895

Boyce BF, Yoneda T, Lowe C, Soriano P, Mundy GR (1992) Requirement of pp60^{c-src} expression for osteoclasts to form ruffled borders and resorb bone in mice. J Clin Invest 90:1622–1627

Boyce BF, Windle JJ, Reddy SV, Wright K, Leach RJ, Roodman GD (1993) Targeting SV-40 T antigen to the osteoclast in transgenic mice causes osteopetrosis, transformation and apoptosis of osteoclasts. J Bone Miner Res 8:S118

Boyce BF, Wright K, Reddy SV, Koop BA, Story B, Devlin R, Leach RJ, Roodman GD, Windle JJ (1995) Targeting simian virus 40 T antigen to the osteoclast in transgenic mice causes osteoclast tumors and transformation and apoptosis of osteoclasts. Endocrinology 136:5751–5759

Bronckers AL, Goei W, Luo G, Karsenty G, D'Souza RN, Lyaruu DM, Burger EH (1996) DNA fragmentation during bone formation in neonatal rodents assessed by transferase-mediated end labeling. J Bone Miner Res 11:1281–1291

Damoulis PD, Hauschka PV (1997) Nitric oxide acts in conjunction with proinflammatory cytokines to promote cell death in osteoblasts. J Bone Miner Res 12:412–422

Dunstan CR, Evans RA, Hills E, Wong SY, Higgs RJ (1990) Bone death in hip fracture in the elderly. Calcif Tissue Int 47:270–275

Dunstan CR, Somers N, Evans R (1993) Osteocyte death and hip fractures. Calcif Tissue Int 53:S113–S117

Elmardi AS, Katchburian MV, Katchburian E (1990) Electron microscopy of developing calvaria reveals images that suggest that osteoclasts engulf and destroy osteocytes during bone resorption. Calcif Tissue Int 46:239–245

Fadok VA, Voelker DR, Campbell PAS, Cohen JJ, Bratton DL, Henson PM (1992) Exposure of phosphatidylserine on the surface of apoptotic lymphocytes triggers specific recognition and removal by macrophages. J Immunol 148:2207–2216

Franzoso G, Carlson L, Xing L, Poljak L, Shores EW, Brown KD, Leonardi A, Tran T, Boyce BF, Siebenlist U (1997) Requirement for NF-κB in osteoclast and B cell development. Genes Dev 11:3482–3496

Fuller K, Owens JM, Jagger CJ, Chambers TJ (1993a) M-CSF suppresses osteoclastic apoptosis and switches function from bone resorption to migration/chemotaxis. J Bone Miner Res 8:S384

Fuller K, Owens JM, Jagger CJ, Wilson A, Moss R, Chambers TJ (1993b) Macrophage colony-stimulating factor stimulates survival and chemotactic behavior in isolated osteoclasts. J Exp Med 178:1733–1744

Gavrieli Y, Sherman Y, Ben-Sasson SA (1992) Identification of programmed cell death in situ via specific labeling of nuclear DNA fragmentation. J Cell Biol 119:493–501

Gold R, Schmied M, Rothe G, Zischler H, Breitschopf H, Wekerle H, Lassmann H (1993) Detection of DNA fragmentation in apoptosis: Application of in situ nick translation to cell culture systems and tissue sections. J Histochem Cytochem 41:1023–1030

Grigoriadis AE, Wang Z-Q, Cecchini MG, Hofstetter W, Felix R, Fleisch HA, Wagner EF (1994) c-Fos: A key regulator of osteoclast-macrophage lineage determination and bone remodeling. Science 266:443–448

Hale AJ, Smith CA, Sutherland LC, Stoneman VEA, Longthorne VL, Culhane AC, Williams GT (1996) Apoptosis: molecular regulation of cell death. Eur J Biochem 236:1–26

Hayman AR, Jones SJ, Boyde A, Foster D, Colledge WH, Carlton MB, Evans MJ, Cox TM (1996) Mice lacking tartrate-resistant acid phosphatase (Acp 5) have disrupted endochondral ossification and mild osteopetrosis. Development 122:3151–3162

Hoffman B, Liebermann DA (1994) Molecular controls of apoptosis: differentiation/growth arrest primary response genes, proto-oncogenes, and tumor suppressor genes as positive and negative modulators. Oncogene 9:1807–1812

Horowitz MC (1993) Cytokines and estrogen in bone; anti-osteoporotic effects. Science 260:626–627

Hughes DE, Wright KR, Mundy GR, Boyce BF (1994a) TGF-β_1 induces osteoclast apoptosis in vitro. J Bone Miner Res 9:S138

Hughes DE, Wright KR, Sasaki A, Yoneda T, Uy H, Roodman GD, Mundy GR, Boyce BF (1994b) Bisphosphonates induce osteoclast apoptosis in vivo and in vitro, but calcitonin does not. J Bone Miner Res 9:S347

Hughes DE, Jilka R, Manolagas S, Dallas SL, Bonewald LF, Mundy GR, Boyce BF (1995a) Sex steroids promote osteoclast apoptosis in vitro and in vivo. J Bone Miner Res 10:S150

Hughes DE, Wright KR, Uy HL, Sasaki A, Yoneda T, Roodman GD, Mundy GR, Boyce BF (1995b) Bisphosphonates promote apoptosis in murine osteoclasts in vitro and in vivo. J Bone Miner Res 10:1478–1487

Hughes DE, Dai A, Tiffee JC, Li HH, Mundy GR, Boyce BF (1996) Estrogen promotes apoptosis of murine osteoclasts mediated by TGF-β. Nature Med 2:1132–1136

Hughes DE, Luckman SP, Graham R, Russell G, Rogers MJ (1997) Involvement of the mevalonate pathway in osteoclast apoptosis and the mechanism of action of bisphosphonates. Bone 20:110S

Iotsova V, Caamano J, Loy J, Lewin A, Bravo R (1997) Osteopetrosis in mice lacking NF-κB1 and NF-κB2. Nature Med 3:1285–1289

Jilka RL, Weinstein RS, Bellido T, Parfitt AM, Manolagas SC (1998) Osteoblast programmed cell death (apoptosis): modulation by growth factors and cytokines. J Bone Miner Res 13:793–802

Jimi E, Shuto T, Koga T (1995) Macrophage colony-stimulating factor and interleukin-1α maintain the survival of osteoclast-like cells. Endocrinology 136:808–811

Jimi E, Ikebe T, Takahashi N, Hirata M, Suda T, Koga T (1996) Interleukin-1α activates an NF-κB like factor in osteoclast-like cells. J Biol Chem 271:4605–4608

Kameda T, Ishikawa H, Tsutsui T (1995) Detection and characterization of apoptosis in osteoclasts in vitro. Biochem Biophys Res Comm 207:753–760

Kameda T, Miyazawa K, Yoshihisa M (1996) Vitamin K_2 inhibits osteoclastic bone resorbtion by inducing osteoclast apoptosis. Biochem Biophys Res Comm 220:515–519

Karaplis AC, Luz A, Glowacki J, Bronson RT, Tybulewicz VL, Kronenberg HM, Mulligan RC (1994) Lethal skeletal dysplasia from targeted disruption of the parathyroid hormone-related peptide gene. Genes Dev 8:277–289

Kerr JFR, Wyllie AH, Currie AR (1972) Apoptosis: A basic biological phenomenon with wide-ranging implications in tissue kinetics. Br J Cancer 26:239–257

Kitajima I, Nakajima T, Imamura T (1996) Induction of apoptosis in murine clonal osteoblasts expressed by human T-cell leukemia virus type I tax by NF-κB and TNF-α. J Bone Miner Res 11:200–210

Lanske B, Karaplis AC, Lee K, Luz A, Vortkamp A, Pirro A, Karperien M, Defize LHK, HO C, Mulligan RC, Abou-Samra AB, Juppner H, Segre GV, Kronenberg HM (1996) PTH/PTHrP receptor in early development and Indian hedgehog-regulated bone growth. Science 273:663–666

Lanyon LE (1993) Osteocytes, strain detection, bone modelling and remodelling. Calcif Tissue Int 53:S102-S107

Lee FD (1993) Importance of apoptosis in the histopathology of drug related lesions in the large intestine. J Clin Pathol 46:118–122

Lee JM, Bernstein A (1993) p53 mutations increase resistance to ionizing radiation. Proc Natl Acad Sci USA 90:5742–5746

Lee K, Lanske B, Karaplis AC, Deeds JD, Kohno H, Nissenson RA, Kronenberg HM, Segre GV (1996) Parathyroid hormone-related peptide delays terminal differentiation of chondrocytes during endochondral bone development. Endocrinology 137:5109–5118

Lewinson D, Silbermann M (1992) Chondroclasts and endothelial cells collaborate in the process of cartilage resorption. Anat Rec 233:504–514

Lowe SW, Ruley HE, Jacks T, Housman DE (1993) p53-dependent apoptosis modulates the cytotoxicity of anticancer agents. Cell 74:957–967

Luckman SP, Hughes DE, Coxon FP, Graham R, Russell G, Rogers MJ (1998) Nitrogen-containing bisphosphonates inhibit the mevalonte pathway and prevent post-translational prenylation of GTP-binding proteins, including Ras. J Bone Miner Res 13:581–589

Lutton JD, Moonga BS, Dempster DW (1996) Osteoclast demise in the rat: physiological versus degenerative cell death. Exp Physiol 81:251–260

Martin SJ, Reutelingsperger CPM, McGahon AJ (1995) Early redistribution of plasma membrane phosphatidylserine is a general feature of apoptosis regardless of the initiating stimulus. Inhibition of overexpression of Bcl-2 and Abl. J Exp Med 182:1545–1557

Mooney EE, Ruis Peris JM, O'Neill A, Sweeney EC (1995) Apoptotic and mitotic indices in malignant melanoma and basal cell carcinoma. J Clin Pathol 48:242–244

Mullender MG, van der Meer DD, Huiskes R, Lips P (1996) Osteocyte density changes in aging and osteoporosis. Bone 18:109–113

Apoptosis in Bone Cells   81

Ozaki K, Takeda H, Iwahashi H, Kitano S, Hanazawa S (1997) NF-κB inhibitors stimulate apoptosis of rabbit mature osteoclasts and inhibit bone resorption by these cells. FEBS Letters 410:297–300

Pacifici R (1996) Estrogen, cytokines, and pathogenesis of postmenopausal osteoporosis. J Bone Miner Res 11:1043–1051

Parfitt AM (1994) Osteonal and hemi-osteonal remodeling: The spatial and temporal framework for signal traffic in adult human bone. J Cell Biochem 55:273–286

Patel T, Gores GJ, Kaufman SH (1996) The role of proteases during apoptosis. FASEB J 10:587–597

Popoff SN, Marks SCJ (1995) The heterogeneity of the osteopetroses reflects the diversity of cellular influences during skeletal development. Bone 17:437–445

Reed JC (1994) Bcl-2 and the regulation of programmed cell death. J Cell Biol 124:1–6

Roach HI, Erenpreisa J, Aigner T (1995) Osteogenic differentiation of hypertrophic chondrocytes involves cell divisions and apoptosis. J Cell Biol 131:483–494

Schwartzberg PL, Xing L, Hoffmann O, Lowell CA, Garrett L, Boyce BF, Varmus HE (1997) Rescue of osteoclast function by transgenic expression of kinase-deficient src in src-/- mutant mice. Genes Dev 11: 2835–2844

Selander KS, Harkonen PL, Valve E, Monkkonen J, Hannuniemi R, Vaananen HK (1996) Calcitonin promotes osteoclast survival in vitro. Mol Cell Endocrinol 122:119–129

Siebenlist U, Franzoso G, Brown K (1994) Structure, regulation and function of NF-κB. Annu Rev Cell Biol 10:405–455

Soriano P, Montgomery C, Geske R, Bradley A (1991) Targeted disruption of the c-src proto-oncogene leads to osteopetrosis in mice. Cell 64:693–702

Tomkinson A, Reeve J, Shaw R, Noble B (1997a) The death of osteocytes via apoptosis accompanies estrogen withdrawal in human bone. J Clin Endocrinol Metab 82:3128–3135

Tomkinson A, Gevers E, Reeve J, Noble BS (1997b) The role of estrogen in the control of osteocyte apoptosis. Bone 20:12S

Tondravi MM, McKercher SR, Anderson K, Erdmann JM, Quiroz M, Maki R, Teitelbaum SL (1997) Osteopetrosis in mice lacking haematopoietic transcription factor PU.1. Nature 386:81–84

Vermes I, Haanen C, Reutelingsperger CPM (1995) A novel assay for apoptosis: Flow cytometric detection of phopshatidylserine expression on early apoptotic cells using fluorescein labeled Annexin V. J Immunol Meth 184:39–51

Villa P, Kaufmann SH, Earnshaw WC (1997) Caspases and caspase inhibitors. Trends Biochem Sci 22: 388–93

Vortkamp A, Lee K, Lanske B, Segre GV, Kronenberg HM, Tabin CJ (1996) Regulation of rate of cartilage differentiation by Indian Hedgehog and PTH-related protein. Science 273:613–621

Wang Z-Q, OVitt C, Grigoriadis AE, Mohle-Steinlein U, Ruther U, Wagner EF (1992) Bone and haematopoietic defects in mice lacking c-Fos. Nature 360:741–745

Weir EC, Philbrick WM, Amling M, Neff LA, Baron R, Broadus AE (1996) Targeted overexpression of parathyroid hormone-related peptide in chondrocytes causes chondrodysplasia and delayed endochondral bone formation. Pro Natl Acad Sci USA 93:10240–10245

White E (1996) Life, death, and the pursuit of apoptosis. Genes Dev 10:1–15

Wright KR, Story B, Hughes DE, Windle J, Reddi S, Roodman GD, Mundy GR, Boyce BF (1995) Standard morphology is more sensitive than TUNEL for identification of apoptosis in osteoclasts. J Bone Miner Res 10:S324

Wyllie AH (1980) Glucocorticoid-induced thymocyte apoptosis is associated with endogenous endonuclease activation. Nature 284:555–556

Wyllie AH, Kerr JFR, Currie AR (1980) Cell death: The significance of apoptosis. Int Rev Cytol 68:251–306

Xing L, Schwartzberg P, Reddy SV, Roodman GD, Mundy GR, Varmus HE, Boyce BF (1996) Induction of osteoclast apoptosis in transgenic mice by a truncated SRC protein. J Bone Miner Res 11:S140

Xing L, Schwartzberg P, Sawyer T, Varmus HE, Boyce BF (1997) Induction of osteoclast apoptosis by mutated *src* proteins and a *src* SH2 inhibitor. J Bone Miner Res 12:S109

Yokouchi Y, Sakiyama J, Kameda T, Iba H, Suzuki A, Ueno N, Kuroiwa A (1996) BMP-2/-4 mediate programmed cell death in chicken limb buds. Development 122:3725–3734

Zou H, Niswander L (1996) Requirement for BMP signaling in interdigital apoptosis and scale formation. Science 272:738–41

4 Genetics in Osteoporosis

B.L. Langdahl and E.F. Eriksen

4.1 Introduction

Low bone mass is a major risk factor for developing osteoporosis (Kanis et al. 1994; Kröger et al. 1995). Several family studies have shown significant correlations between bone mass in pre-menopausal sisters (Hansen et al. 1992) and in mothers and daughters (Hansen et al. 1992; Lutz and Tesar 1990; Kahn et al. 1994). A study comprising 16 families revealed that bone mass in teenagers was significantly correlated to both their mothers and fathers bone mass, but with the highest correlation coefficient to the mean of their mothers and fathers bone mass (Lonzer et al. 1996). Two Australian studies have demonstrated reduced bone mass compared to age-matched normal controls in daughters of women with spinal osteoporosis or fracture of the femoral neck (Seeman et al.

1994a, b). In a study comprising a large population from California, both paternal and maternal histories of osteoporosis were shown to be risk factors for osteoporosis. In fact, the relative risk of osteopenia was higher in individuals with a father with osteoporosis: 2.16 compared to individuals with a mother suffering from osteoporosis (Soroko et al. 1994).

Several twin studies have confirmed these results and estimated that genetic factors are responsible for up to 75% of the inter-individual variation in bone mass (Arden et al. 1996; Kelly et al. 1991; Pocock et al. 1987; Hustmyer et al. 1994). Most of these studies referred to peak bone mass, but bone loss has also been demonstrated to be influenced by genetic factors (Garnero et al. 1996; Kelly et al. 1993).

Segregation analysis in families has shown that bone mass is likely to be under the control of several genes with modest effects, rather than a small number of genes with large effects (Soroko et al. 1994; Guegen et al. 1995; Sowers et al. 1992). Despite the fact that much effort has been spent in many research laboratories over the last several years, the genes involved have not yet been sufficiently characterized.

4.2 Vitamin D receptor

The first polymorphism associated with reduced bone mass was a poly-morphism in an untranslated region of the vitamin D receptor (VDR) gene. Morrison et al. (1992) found a highly significant association between this polymorphism and serum levels of osteocalcin. Later on, the same group showed highly significant correlations between the VDR genotypes and bone mass and serum levels of vitamin D (Morrison et al. 1994). The results in the twins were, however, later on retracted (Morrison et al. 1997). It is still not clear how this polymorphism affects production of the vitamin D receptor. Expression studies showed in-creased levels of mRNA in cells with the *BB* genotype, associated with reduced bone mass and increased levels of vitamin D (Generally, poly-morphisms are named according to the restriction enzyme cleaving site they generate or remove, in VDR the three polymorphisms are identified by *Bsm*I, *Taq*I or *Apa*I. Capitalized letters stands for the absence of and lower case for the presence of the restriction site.) Following this initial study, many studies have been conducted in order to confirm these results and further clarify the underlying mechanisms. Some studies confirmed the data (Fleet et al. 1995; Howard et al. 1995; Riggs et al. 1995; Spector et al. 1995; Tokita et al. 1996; Viitanen et al. 1996; Sigurdsson et al. 1997; Kiel et al. 1997), some could not find any associations (Hustmyer et al. 1994; Riggs et al. 1995; Kiel et al. 1997; Franols et al. 1997; Keen et al. 1995; Lim et al. 1995; Jørgensen et al. 1996; Spotila et al. 1996;Alahari et al. 1997; Tsai et al. 1996; Berg et al. 1996; Melhus et al. 1994; Rauch et al. 1997; Zmuda et al. 1997) and some even found the reverse association (Houston et al. 1996; Uitterlinden et al. 1996). Furthermore, two studies demonstrated increased bone loss in post-menopausal women with the *BB* genotype (Krall et al. 1995; Ferrari et al. 1995), but two other studies showed no associations between the VDR genotypes and bone loss (Rauch et al. 1997; Garnero et al. 1996). If this polymorphism of the VDR affected bone mass in a significant way, it was to be expected that genotype distribution was different in osteoporotic and normal populations. This has not been found in any studies (Melhus et al. 1994; Tamai et al. 1997; Vandevyver et al. 1997; Looney et al. 1995; Langdahl et al. 1996) and has led to the hypothesis that it is not the VDR gene locus itself that is related to bone mass, but it is perhaps linked to a neighboring gene that affects bone metabolism (linkage disequilibrium).

Dawson-Hughes and colleagues demonstrated that the effect of the VDR polymorphism was influenced by environment. Intestinal calcium absorption was unaffected by VDR genotype in women on high calcium intake, whereas calcium absorption was reduced in women with the *BB* genotype compared to women with the bb genotype during a period with low calcium intake (Dawson-Hughes et al. 1995). This modifying effect of calcium intake was confirmed in studies comprising 229 American and 101 African-American post-menopausal women (Zmuda et al. 1997; Krall et al. 1995); however, the data could not be confirmed in a small UK study (Francis et al. 1997). In a Danish case control study comprising 139 osteoporotic patients with vertebral fractures and normal controls, we could not demonstrate any differences in serum levels of 25-hydroxy vitamin D and 1,25-dihydroxy vitamin D or in urinary excretion of calcium between the genotypes (Langdahl et al. 1996). In a Dutch study it was demonstrated that the effect of vitamin D supplementation in elderly women is dependent on VDR genotypes. In this study only women with the *BB* or the *Bb* genotypes responded with an increase in bone mineral density (BMD) of the femoral neck, whereas no significant changes in bone mass were detected in women with the bb genotype (Graafmans et al. 1997). Recently, a significant correlation between VDR genotypes and bone mass of the lumbar spine and the femoral neck was demonstrated in prepubertal American girls of Mexican descent (Sainz et al. 1997). However, these correlations could not be found in Norwegian children and adolescents. Furthermore, no associations could be demonstrated between the VDR genotypes and gain of bone in these children (Gunnes et al. 1997).

At the other end of the VDR gene a polymorphism that affects the transcription initiation site in the second exon has been found. This polymorphism, a substitution of thymine with cytosine, deletes the normal transcription initiation site and transcription is initiated at the normal codon 4. The effect of this polymorphism on bone mass has been examined in three studies so far. In a study comprising 100 normal Mexican-American women 15% had the *ff* genotype and significantly reduced bone mass at the lumbar spine (Gross et al. 1996). Harris et al. reported that the *ff* genotype is only found in 4% of black American women, which is significantly different from the prevalence in white American women. In this study only associations with bone mass at the hip could be demonstrated (Harris et al. 1997). In a Japanese study the genotype distribution was similar with the distribution in the white American population, and the difference in bone mass at the lumbar spine between the genotypes was also around 10% and significant (Arai et al. 1997). This study also demonstrated reduced responsiveness of the variant VDR in HeLa cells transfected with a VDR expression vector and a luciferase reporter construct compared to the wild-type VDR.

4.3 Transforming Growth Factor-β1

Transforming growth factor-β1(TGF-β1) has been assigned a putative important role in controlling the coupling between osteoclastic bone resorption and osteoblastic bone formation. TGF-β1 is produced by the osteoblasts and laid down in the new formed bone matrix, where it has been found in high concentrations (Seyedin et al. 1986). During bone resorption TGF-β1 is released and activated (Oreffo et al. 1989) and reduces further resorptive activity (Chenu et al. 1988) as well as stimulating osteoblasts (Pfeilschifter et al. 1987). We examined the entire coding region of the TGF-β1 gene using the single-stranded conformation polymorphism method (SSCP) and sequencing. We examined 161 women with osteoporotic vertebral fractures and 131 normal controls and found two variations in the sequence. A cytosine to thymine base substitution in exon 5 was found with the same frequency in osteoporotic patients as in normal controls and did not affect bone mass. In the intron sequence leading up to exon 5 we found a deletion of a cytosine in position –8. This deletion was significantly more prevalent in osteoporotic patients than in normal controls (χ^2=4.02, p<0.05) and was associated with reduced bone mass in osteoporotic women and increased bone turnover, as evaluated by biochemical markers, in both osteoporotic and normal women (Fig. 1) (Langdahl et al. 1997a).

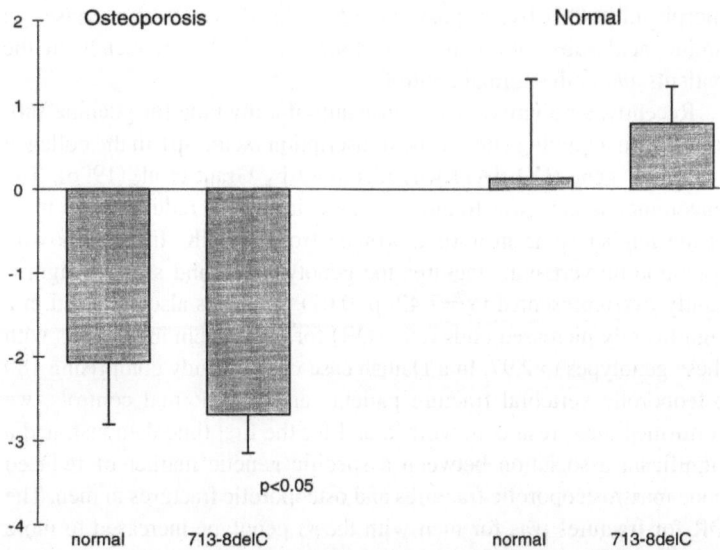

Fig. 1. Bone mineral density (Z-score) of the lumbar spine in osteoporotic patients and normal controls with and without a deletion in the TGF-β1 gene (713–8delC)

4.4 Collagen

Type I collagen constitutes more than 90% of bone, and defects in the production or excessive removal of collagen could be implicated in the pathogenesis of osteoporosis. The more extreme form of osteoporosis, osteogenesis imperfecta, is caused by mutations in the collagen genes. The genes encoding collagen are therefore obvious candidate genes in osteoporosis. Thiry-Blaise et al. (1995) reported three variations in the collagen type I gene upstream from position −300. They could not find any associations between these variations and osteoporosis. Spotila et al. (1994) sequenced both COL1A1 and COL1A2 in 26 patients with osteoporosis, 37 with osteopenia and 81 normal controls. They found three patients with altered sequence, one with a formerly described polymorphism, and two with a new sequence variation that changes an amino acid from proline to alanine. In addition, they described a polymorphism in COLIA2 at position −459. This polymorphism causes an amino acid substitution, but was found in similar frequency in the patients and in the normal controls.

Recently, a polymorphism consisting of a thymine for guanine substitution in a binding site for the transcription factor Sp1 in the collagen type I α1 gene (COLIA1) was identified by Grant et al. (1996). The uncommon allele s was found to be associated with reduced bone mass at the lumbar spine in normal women from the UK. In patients with osteoporotic vertebral fractures the genotypes Ss and ss were significantly overrepresented (χ^2=7.42, p<0.01). This was also reflected in a significantly increased odds ratio (OR) for fractures in individuals with these genotypes to 2.97. In a Danish case-control study comprising 180 osteoporotic vertebral fracture patients and 195 normal controls, we confirmed these results in women and for the first time demonstrated a significant association between a specific genetic marker of reduced bone mass/osteoporotic fractures and osteoporotic fractures in men. The OR for fractures was for men with the ss genotype increased to more than 15, and in men with either Ss or ss genotypes the OR was doubled (Langdahl et al. 1998) (Fig. 2). Despite these significant increases in

→

Fig. 2. Prevalences of the COLIA1 genotypes in osteoporotic (*OP*) patients and normal (*N*) controls

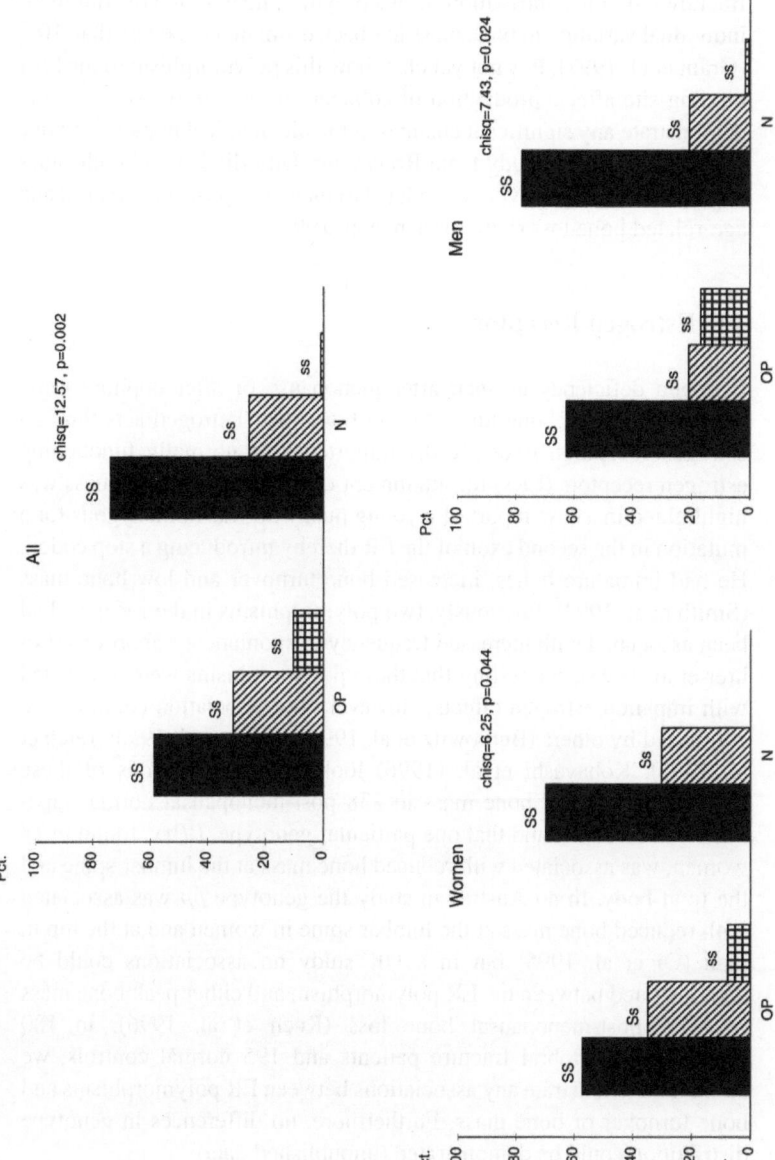

fracture risk, the contribution of this polymorphism to the overall inter-individual variation in bone mass has been estimated to be less than 10% (Grant et al. 1997). It is not yet clear how this polymorphism in the Sp 1 binding site affects production of collagen. In our study, we could not demonstrate any significant changes in the biochemical markers of bone turnover. In a large study from Rotterdam, Uitterlinden and colleagues found that the *ss* genotype was related to increased post-menopausal and age-related bone loss (Uitterlinden et al. 1997).

4.5 Estrogen Receptor

Estrogen deficiency as seen after menopause or after oophorectomy induces increased bone turnover and bone loss. Estrogen acts through estrogen receptors. Recently the importance of normally functioning estrogen receptors (ERs) for attainment of normal peak bone mass was highlighted in a case report of a young man who was homozygous for a mutation in the second exon of the ER thereby introducing a stop codon. He had immature bones, increased bone turnover and low bone mass (Smith et al. 1994). Previously, two polymorphisms in the ER gene had been associated with increased frequency of spontaneous abortions (Le-hrer et al. 1993), suggesting that these polymorphisms were associated with impaired estrogen effects. However, this association could not be confirmed by others (Berkowitz et al. 1994; Taylor et al. 1993; Yaich et al. 1992). Kobayashi et al. (1996) looked for associations of these polymorphisms and bone mass in 238 post-menopausal normal Japanese women and found that one particular genotype, *PPxx*, found in 18 women, was associated with reduced bone mass at the lumbar spine and the total body. In an Australian study the genotype *pp* was associated with reduced bone mass at the lumbar spine in women and at the hip in men (Oi et al. 1995) but in a UK study no associations could be demonstrated between the ER polymorphism and either peak bone mass or early post-menopausal bone loss (Keen et al. 1996). In 180 osteoporotic vertebral fracture patients and 195 normal controls, we could not demonstrate any associations between ER polymorphisms and bone turnover or bone mass. Furthermore, no differences in genotype distribution could be demonstrated (unpublished data).

In another Japanese study by Sano et al. (1995) a TA repeat region upstream from the ER was examined. The number of repeats varied between ten and 27. Women with 12 repeats had reduced bone mass at the lumbar spine and increased levels of biochemical markers of bone turnover. However, there was no general correlation between number of repeats and bone mass or bone turnover.

4.6 Interleukin-1β and Interleukin-1 Receptor Antagonist

Interleukin-1β (IL-1β) is a potent stimulator of bone resorption itself (Canilis et al. 1988; Linkhart and MacCharles 1992; Thomson et al. 1986) and stimulates the production of IL-6, which also has been shown to stimulate bone resorption in rodent and human osteoblasts (Girasol et al. 1992; Passeri et al. 1993). The effect of IL-1β is competitively inhibited by the interleukin-receptor antagonist (IL-1ra), which binds to the IL-1 receptor without inducing any effect. It is therefore the ratio IL-1β/IL-1ra that is responsible for the biological effects and an imbalance in this ratio has been associated with post-menopausal bone loss (Khosla et al. 1994; Kimble et al. 1995; Ralston 1994). We examined the entire coding region of the IL-1β gene and found a sequence variation without effect on the amino acid sequence in two normal women. A polymorphism has been demonstrated in the promoter of IL-1β (diGiovines et al. 1992). We found no effect of this polymorphism on bone mass or bone turnover in 378 osteoporotic patients and normal controls. A polymorphism in the IL-1ra gene, consisting of a variable number of repeats of an 86 base pair sequence, has been shown to affect leukocyte production of IL-1ra (Danis et al. 1995). We found, in 185 osteoporotic patients with vertebral fractures and 199 normal controls, that the genotypes previously demonstrated to be associated with decreased production of IL-1ra, were significantly more frequent in osteoporotic patients (56.2%) than in age matched normal controls (43.3%) (χ^2=3.85, p<0.05). The relative risk of osteoporotic fractures was significantly increased to 1.68. In the same individuals, there was a trend towards decreased bone mass, and the difference increased with increasing age (Langdahl et al. 1997b). This last finding is in accordance with results from a study on early post-menopausal bone loss by Woodford-Richens et al. (1997), demonstrating increased rate of bone loss in women with the genotypes associated with reduced production of IL-1ra.

4.7 Interleukin-6

Interleukin-6 (IL-6) is a potent stimulator of bone resorption and has been implicated in the pathogenesis of bone loss associated with estrogen deficiency (Girasole et al. 1992; Passeri et al. 1993). Murray and colleagues have demonstrated an association between a polymorphic TA repeat at the 3' flank of the IL-6 gene and bone mass in 153 normal Scottish women (Murray et al. 1997). Woodford-Richens and colleagues did not find any effect of this polymorphism on either peak bone mass or early post-menopausal bone loss in normal British women (Woodford-Richens et al. 1997). We have examined the coding region of the IL-6 gene and found two sequence variations both without effects on the amino acid sequences. We found eight osteoporotic women with a substitution of thymine for guanine at position –62 in exon 3 and four osteoporotic and seven normal women with a substitution of thymine for cytosine in exon 5. Both sequence variations were without effect on bone mass or bone turnover (unpublished data).

4.8 Insulin Like Growth Factor-I

Insulin like growth factor-I (IGF-I) is a potent stimulator of osteoblastic proliferation. It has been demonstrated that osteoporotic patients not only have increased resorptive activity but also display impaired bone formation (Eriksen et al. 1990). Furthermore, patients with low levels of circulating IGF-I have reduced bone forming capability (Johansson et al. 1997) and that serum levels of IGF-I are reduced in women with established osteoporosis compared to normal age-matched women (Ravn et al. 1995). We examined the entire coding region of the IGF-I gene and surrounding introns and found a substitution in the sequence leading up to the first exon of adenine with cytosine. The sequence variation is rare, found in 8% of both osteoporotic and normal women, and without effect on bone mass or bone turnover (unpublished data).

4.9 Apolipoprotein E

Apolipoprotein E (Apo E) is a major component of high-density and low-density lipoproteins. There are three principal isoforms which previously were reported to be related to involutional diseases, such as Alzheimers dementia and cardiovascular risk in diabetes mellitus. Moreover, serum levels of vitamin K, an activator of osteocalcin, have been related to Apo E phenotype. Recently, Shiraki et al. (1997) demonstrated an association between phenotypes of Apo E and bone mass in the lumbar spine and total body in Japanese post-menopausal women. Serum levels of osteocalcin were higher in the phenotypes associated with reduced bone mass; however, no correlations were found with urinary excretion of pyridinoline and deoxypyridinoline.

4.10 Future Aspects

In the years to come, many more candidate genes will be examined. Linkage analysis studies comprising sib-pairs and parents and children will identify more genes involved in the pathogenesis of osteoporosis. In addition, more attention will probably be given to gene-gene and gene-environment interactions.

We think that the polymorphisms associated with increased fracture risk or decreased bone mass, in the near future, will be incorporated in the evaluation of the risk of developing osteoporosis in the individual. Most likely, we will end up with an array of genes to be considered.

References

Alahari KD, Lobaugh B, Econs MJ (1997) Vitamin D receptor alleles do not correlate with bone mineral density in premenopausal Caucasian women from the southeastern United States. Metabolism 46: 224–226

Arai H, Miyamoto K-I, Taketani Y, Yamamoto H, Iemori Y, Morita K, Tonai T, Nishisho T, Mori S, Takeda E (1997) A vitamin D receptor gene polymorphism in the translation initiation codon: Effect on protein activity and relation to bone mineral density in Japanese women. J Bone Miner Res 12: 915–921

Arden NK, Baker J, Hogg C, Baan K, Spector TD (1996) The heritability of bone mineral density, ultrasound of the calcaneus and hip axis length: a study og postmenopausal twins. J Bone Miner Res 11: 530–534

Berg JP, Falch JA, Haug E (1996) Fracture rate, pre and postmenopausal bone mass and early and late postmenopausal bone loss are not associated with vitamin D receptor genotype in a high – endemic area of osteoporosis. Eur J Endocrinol 135: 96–100

Berkowitz GS, Stone JL, Lehrer SP, Marcus M, Lapinski RH, Schachter BS (1994) An estrogen receptor genetic polymorphism and the risk of primary and secondary recurrent spontaneous abortion. Am J Obstet Gynecol 171: 1579–1584

Canalis E, McCarthy T, Centrella M (1988) Growth factors and the regulation of bone remodeling. J Clin Invest 81: 227–281

Chenu C, Pfeilschifter J, Mundy GR, Roodman GD (1988) Transforming growth factor b inhibits formation of osteoclast – like cells in long-term human marrow cultures. Proc Natl Acad Sci USA 85: 5683–5687

Danis VA, Millington M, Hyland VJ, Grennan D (1995) Cytokine production by normal human monocytes: inter-subject variation and relationship to an IL-1 receptor antagonist (IL-1Ra) gene polymorphism. Clin Exp Immunol 99: 303–310

Dawson-Hughes B, Harris SS, Finneran S (1995) Calcium absorption on high and low calcium intakes in relation to vitamin D receptor genotype. J Clin Endocrinol Metab 80: 3657–3661

diGiovine FS, Takhsn E, Blakemore AIF, Duff GW (1992) Single base polymorphism at −511 in the human interleukin-1b gene (IL1b). Hum Mol Gen 1: 450

Eriksen EF, Hodgson SF, Eastell R, Cedel SL, O'Fallon WM, Riggs BL (1990) Cancellous bone remodeling in type I (postmenopausal) osteoporosis: Quantitative assessment of rates of formation, resorption, and bone loss at tissue and cellular levels. J Bone Miner Res 5: 311–319

Ferrari S, Rizzoli R, Chevalley T, Slosman D, Eisman JA, Bonjour J-P (1995) Vitamin-D-receptor-gene polymorphisms and change in lumbar-spine bone mineral density. Lancet 345: 423–424

Fleet JC, Harris SS, Wood RJ, Dawson-Hughes B (1995) The BsmI vitamin D receptor restriction fragment length polymorphism (BB) predicts low bone density in premenopausal black and white women. J Bone Miner Res 10: 985–990

Francis RM, Harrington F, Turner E, Papiha SS, Datta HK (1997) Vitamin D receptor gene polymorphism in men and its effect on bone density and calcium absorption. Clin Endocrinol Oxf 46: 83–86

Franols RM, Harrington F, Turner E, Paplha SS, Datta HK (1997) Vitamin D receptor gene polymorphism in men and its effect on bone density and calcium absorption. Clin Endocrinol Oxf 46: 83–86

Garnero P, Arden NK, Griffiths G, Delmas PD, Spector TD (1996) Genetic influence on bone turnover in postmenopausal twins. J Clin Endocrinol Metab 81: 140–146

Garnero P, Borel O, Sornay-Rendu E, Arlot ME, Delmas PD (1996) Vitamin D receptor gene polymorphisms are not related to bone turnover, rate of bone loss, and bone mass in postmenopausal women: the OFELY Study. J Bone Miner Res 11: 827–834

Girasole G, Jilka RL, Passeri G, Boswell S, Boder G, Williams DC, Manolagas SC (1992) 17b-estradiol inhibits interleukin-6 production by bone marrow-derived stromal cells and osteoblasts in vitro: a potential mechanism for the antiosteoporotic effect of estrogens. J Clin Invest 89: 883–891

Graafmans WC, Lips P, Ooms ME, Leeuwen JPTMv, Pols HAP, Uitterlinden AG (1997) The effect of vitamin D supplementation on the bone mineral density of the femoral neck is associated with vitamin D receptor genotype. J Bone Miner Res 12: 1241–1245

Grant SFA, Nguyen TV, Howard GM, White C, Morrison NA, Ralston SH, Eisman JA (1997) Genetic linkage between a polymorphism in the collagen I alpha I gene and bone mineral density: A twin study. Bone 20: 7S

Grant SFA, Reid DM, Blake G, Herd R, Fogelman I, Ralston SH (1996) Reduced bone density and osteoporosis associated with a polymorphic Sp1 binding site in the collagen type I a 1 gene. Nature Gen 14: 203–205

Gross C, Eccleshall TR, Malloy PJ, Villa ML, Marcus R, Feldman D (1996) The presence of a polymorphism af the translation initiation site of the vitamin D receptor gene is associated with low bone mineral density in postmenopausal Mexican-American women. J Bone Miner Res 11: 1850–1855

Gueguen R, Jouanny P, Guillemin F, Kuntz C, Pourel J, Siest G (1995) Segregation analysis and variance components analysis of bone mineral density in healthy families. J Bone Miner Res 12: 2017–2022

Gunnes M, Berg JP, Halse J, Lehmann EH (1997) Lack of relationship between vitamin D receptor genotype and forearm bone gain in healthy children, adolescents, and young adults. J Clin Endocrinol Metab 82: 851–855

Hansen MA, Hassager C, Jensen SB, Christiansen C (1992) Is heritability a risk factor for postmenopausal osteoporosis ? J Bone Miner Res 7: 1037–1043

Harris SS, Eccleshall TR, Gross C, Dawson-Hughes B, Feldman D (1997) The vitamin D receptor start codon polymorphism (FokI) and bone mineral density in premenopausal American black and white women. J Bone Miner Res 12: 1043–1048

Houston LA, Grant SF, Reid DM, Ralston SH (1996) Vitamin D receptor polymorphism, bone mineral density, and osteoporotic vertebral fractures: studies in a UK population. Bone 18: 249–252
Howard G, Nguyen T, Morrison N, Watanabe T, Sambrook P, Eisman J, Kelly
PJ (1995) Genetic influences on bone density: Physiological correlates of
vitamin D receptor gene alleles in premenopausal women. J Clin Endocrinol Metab 80: 2800–2805
Hustmyer FG, Peacock M, Hui S, Johnston C, Christian J (1994) Bone mineral
density in relation to polymorphism at the vitamin D receptor gene locus. J
Clin Invest 94: 2130–2134
Johansson AG, Eriksen EF, Lindh E, Langdahl B, Blum WF, Lindahl A, Ljung-
gren ..., Ljunghall S (1997) Reduced serum levels of the growth hormone-
dependent IGF binding protein and a negative bone balance at the level of
individual remodeling units idiopathic osteoporosis in men. J Clin Endocrinol Metab 82: 2795–2798
Jørgensen HL, Scholler J, Sand JC, Bjurting M, Hassager C, Christiansen C
(1996) Relation of common allelic variation in vitamin D receptor locus to
bone mineral density and postmenopausal bone loss: cross sectional and
longitudinal population study. Br Med J 313: 586–590
Kahn SA, Pace JE, Cox ML, Gau DW, Cox SA, Hodkinson HM (1994)
Osteoporosis and genetic influence: a three-generation study. Postgrad Med
J 70: 798–800
Kanis JA, Melton JR 3rd, Christiansen C, Johnston CC, Khaltaev N (1994)
The diagnosis of osteoporosis. J Bone Miner Res 9: 1137–1141
Keen RW, Major PJ, Lanchbury JS, Spector TD (1995) Vitamin D receptor
gene polymorphisms and bone loss. Lancet 345: 990
Keen RW, Woodford-Richens KL, Lanchbury JS, Spector TD (1996) Peak
bone mass, early postmenopausal bone loss and polymorphism at the estro-
gen receptor gene. Osteoporosis Int 6: 102
Kelly PJ, Hopper JL, Macaskill GT, Pocock NA, Sambrook PN, Eisman JA
(1991) Genetic factors in bone turnover. J Clin Endocrinol Metab 72:
808–813
Kelly PJ, Nguyen T, Hopper J, Pocock N, Sambrook P, Eisman J (1993)
Changes in axial bone density with age: A twin study. J Bone Miner Res 8:
11–17
Khosla S, Peterson JM, Egan K, Jones JD, Riggs BL (1994) Circulating cytokine levels in osteoporotic and normal women. J Clin Endocrinol Metab 79:
707–711
Kiel DP, Myers RH, Cupples La, Kong XF, Zhu XH, Ordovas J, Schaefer EJ,
Felson DT, Rush D, Wilson PWF, Eisman JA, Holick MF (1997) The BsmI
vitamin D receptor restriction fragment length polymorphism (bb) influ-

ences the effect of calcium intake on bone mineral density. J Bone Miner Res 12: 1049–1057

Kimble RB, Matayoshi AB, Vannice JL, Kung VT, Williams C, Pacifici R (1995) Simultaneous block of interleukin-1 and tumor necrosis factor is required to completely prevent bone loss in the early postovariectomy period. Endocrinology 136: 3054–3061

Kobayashi S, Inoue S, Hosoi T, Ouchi Y, Shiraki M, Orimo H (1996) Association of bone mineral density with polymorphism of the estrogen receptor gene. J Bone Miner Res 11: 306–311

Krall EA, Parry P, Lichter JB, Dawson-Hughes B (1995) Vitamin D receptor alleles and rates of bone loss: Influences of years since menopause and calcium intake. J Bone Miner Res 10: 978–984

Kröger H, Huopio J, Honkanen R, Tuppurainen M, Puntila E, Alhava E, Saarikoski S (1995) Prediction of fracture risk using axial bone mineral density in a perimenopausal population: A prospective study. J Bone Miner Res 10: 302–306

Langdahl BL, Gravholt CH, Eriksen EF (1996) No correlations between vitamin D receptor polymorphisms and bone mass and bone turnover in danish osteoporotic and normal women. Osteoporosis Int 6: 153

Langdahl BL, Knudsen JY, Jensen HK, Gregersen N, Eriksen EF (1997) A sequence variation: 713-8delC in the transforming growth factor-b1 gene has higher prevalence in osteoporotic women than in normal women and is associated with very low bone mass in osteoporotic women and increased bone turnover in both osteoporotic and normal women. Bone 20: 289–294

Langdahl BL, Løkke E, Carstens M, Stenkjær LL, Eriksen EF (1997) A 86 base pair repeat polymorphism in the interleukin 1 receptor antagonist gene is associated with osteoporotic fractures, sequence variations in the coding regions and a polymorphism in the promoter ot the interleukin 1b gene are not. submitted :

Langdahl BL, Ralston SH, Grant SFA, Eriksen EF (1998) Collagen type I a 1 gene Sp1 polymorphism predicts bone density and osteoporotic fracture in both men and women. J Bone Miner Res (in press)

Lehrer SP, Schmutzler RK, Rabin JM, Schachter BS (1993) An estrogen receptor genetic polymorphism and a history of spontaneous abortion – correlation in women with estrogen receptor positive breast cancer but not in women with estrogen receptor negative breast cancer or in women without cancer. Breast Cancer Res Treat 26: 175–180

Lim SK, Park YS, Park JM, Song YD, Lee EJ, Kim KR, Lee HC, Huh KB (1995) Lack of association between vitamin D receptor genotypes and osteoporosis in Koreans. J Clin Endocrinol Metab 80: 3677–3681

Linkhart TA, MacCharles DC (1992) Interleukin-1 stimulates release of insulin-like growth factor-I from neonatal mouse calvaria by a prostaglandin synthesis-dependent mechanism. Endocrinology 131: 2297–2305

Lonzer MD, Imrie R, Rogers D, Worley D, Licata A, Secic M (1996) Effects of heredity, age, weight, puberty, activity, and calcium intake on bone mineral density in children. Clin Pediatr Phila 35: 185–189

Looney JE, Yoon HK, Fischer M, Farley SM, Farley JR, Wergedal JE, Baylink DJ (1995) Lack of a high prevalence of the BB vitamin D receptor genotype in severely osteoporotic women. J Clin Endocrinol Metab 80: 2158–2162

Lutz J, Tesar R (1990) Mother-daughter pairs: spinal and femoral bone densities and dietary intakes. Am J Clin Nutr 52: 872–877

Melhus H, Kindmark A, Am r S, Wil n B, Lindh E, Ljunghall S (1994) Vitamin D receptor genotypes in osteoporosis. Lancet 344: 949–950

Morrison NA, Qi JC, Tokita A, Kelly PJ, Crofts L, Nguyen TV, Sambrook PN, Eisman JA (1994) Prediction of bone density from vitamin D receptor alleles. Nature 367: 284–287

Morrison NA, Qi JC, Tokita A, Kelly PJ, Crofts L, Nguyen TV, Sambrook PN, Eisman JA (1997) Correction: Prediction of bone density from vitamin D receptor alleles. Nature 387: 106

Morrison NA, Yeoman R, Kelly PJ, Eisman JA (1992) Contribution of trans-acting factor alleles to normal physiological variability: Vitamin D receptor gene polymorphisms and circulating osteocalcin. Proc Natl Acad Sci USA 89: 6665–6669

Murray RE, McGuigan F, Grant SFA, Reid DM, Ralston SH (1997) Polymorphisms of the interleukin-6 gene are associated with bone mineral density. Bone 21: 89–92

Oi JC, Morrison NA, Nguyen TV, White CP, Kelly PJ, Sambrook PN, Eisman JA (1995) Estrogen receptor genotypes and bone mineral density in women and men. J Bone Miner Res 10: S170

Oreffo ROC, Bonewald L, Kukita A, Garrett IR, Seyedin SM, Rosen D, Mundy GR (1990) Inhibitory effects of the bone-derived growth factors osteoinductive factor and transforming growth factor-β on isolated osteoclasts. Endocrinology 126: 3069–3075

Oreffo ROC, Mundy GR, Seyedin SM, Bonewald LF (1989) Activation of the bone-derived latent TGF beta complex by isolated osteoclasts. Biochem Biophys Res Comm 158: 817–823

Pacifici R, Rifas L, Teitelbaum S, Slatopolsky E, McCracken R, Bergfeld M, Lee W, Avioli LV, Peck WA (1987) Spontaneous release of interleukin 1 from human blood monocytes reflects bone formation in indiopathic osteoporosis. Proc Natl Acad Sci USA 84: 4616–4620

Passeri G, Girasole G, Jilka RL, Manolagas SC (1993) Increased interleukin-6 production by murine bone marrow and bone cells after estrogen withdrawal. Endocrinology 133: 822–828

Pfeilschifter J, D'Souza SM, Mundy GR (1987) Effects of transforming growth factor-b on osteoblastic osteosarcoma cells. Endocrinology 121: 212–218

Pocock NA, Eisman JA, Hopper JL, Yeates MG, Sambrook PN, Eberi S (1987) Genetic deteminants of bone mass in adults. A twin study. J Clin Invest 80: 706–710

Ralston SH (1994) Analysis of gene expression in human bone biopsies by polymerase chain reaction: Evidence for enhanced cytokine expression in postmenopausal osteoporosis. J Bone Miner Res 9: 883–890

Rauch R, Radermacher A, Danz A, Schiedermaier U, GolŸcke A, Michalk D, Schšnau E (1997) Vitamin D receptor genotypes and changes of bone density in physically active German women with high calcium intake. Exp Clin Endocrinol Diabetes 105: 103–108

Ravn P, Overgaard K, Spencer EM, Christiansen C (1995) Insulin-like growth factors I and II in healthy women with and without established osteoporosis. Eur J Endocrinol 132: 313–319

Riggs BL, Nguyen TV, III LJM, Morrison NA, O'Fallon WM, Kelly PJ, Egan KS, Sambrook PN, Muhs JM, Eisman JA (1995) The contribution of vitamin D receptor gene alleles to the determination of bone mineral density in normal and osteoporotic women. J Bone Miner Res 10: 991–996

Sainz J, Tornout JMv, Loro ML, Sayre J, Roe TF, Gilsanz V (1997) Vitamin D-receptor gene polymorphisms and bone density in prepubertal american girls of Mexican descent. N Engl J Med 337: 77–82

Sano M, Inoue S, Hosoi T, Ouchi Y, Emi M, Shiraki M, Orimo H (1995) Association of estrogen receptor dinucleotide repeat polymorphism with osteoporosis. Biochem Biophys Res Comm 217: 378–383

Seeman E (1994) Reduced bone density in women with fractures: contribution of low peak bone density and rapid bone loss. Osteoporosis Int 4: 15–25

Seeman E, Tsalamandris C, Formica C, Hopper JL, McKay J (1994) Reduced femoral neck bone density in the daughters of women with hip fractures: the role of low peak bone density in the pathogenesis of osteoporosis. J Bone Miner Res 9: 739–743

Seyedin SM, Thompson AY, Bentz H, Rosen DM, McPherson JM, Conti A, Siegel NR, Galluppi GR, Piez KA (1986) Cartilage-inducing factor-A. J Biol Chem 261: 5693–5695

Shiraki M, Shiraki Y, Aoki C, Hosoi T, Inoue S, Kaneki M, Ouchi Y (1997) Association of bone mineral density with apolipoprotein E phenotype. J Bone Miner Res 12: 1438–1445

Sigurdsson G, Magnusdottir DN, Kristinsson J..., Kristjansson K, Olafsson I (1997) Association of BsmI vitamin D receptor gene polymorphism with combined bone mass in spine and proximal femur in Icelandic women. J Int Med 241: 501–505

Smith EP, Boyd J, Frank GR, Takahashi H, Cohen RM, Specker B, Williams TC, Lubahn DB, Korach KS (1994) Estrogen resistance caused by a mutation in the estrogen-receptor gene in a man. N Engl J Med 331: 1056–1061

Soroko SB, Barret-Connor E, Edelstein SL, Kritz-Silverstein D (1994) Family history of osteoporosis and bone mineral density at the axial skeleton: The Rancho Bernardo study. J Bone Miner Res 9: 761–769

Sowers MR, Boehnke M, Jannausch ML, Crutchfield M, Corton G, Burns TL (1992) Familiality and partitioning the variability of femoral bone mineral density in women of child -bearing age. Calcif Tissue Int 50: 110–114

Spector TD, Keen RW, Arden NK, Morrison NA, Major PJ, Nguyen TV, Kelly PJ, Baker JR, Sambrook PN, Lanchbury JS, Eisman JA (1995) Influence of vitamin D receptor genotype on bone mineral density in postmenopausal women: a twin study in Britain. Br Med J 310: 1357–1360

Spotila LD, Caminis J, Johnston R, Shimoya KS, O'Connor MP, Prockop DJ, Tenenhouse A, Tenenhouse HS (1996) Vitamin D receptor genotype is not associated with bone mineral density in three etnic/regional groups. Calcif Tissue Int 59: 235–237

Spotila LD, Colige A, Sereda L, Constantinou-Deltas CD, White MP, Riggs BL, Shaker JL, Spector TD, Hume E, Olsen N, Attie M, Tenenhouse A, Shane E, Briney W, Prockop DJ (1994) Mutation analysis of coding sequences for type I procollagen in individuals with low bone density. J Bone Miner Res 9: 923–932

Tamai M, Yokouchi M, Komiya S, Mochizuki K, Hidaka S, Narita S, Inoue A, Itoh K (1997) Correlation between vitamin D receptor genotypes and bone mineral density in Japanese patients with osteoporosis. Calcif Tissue Int 60: 229–232

Taylor JA, Wilcox AJ, Bowes WA, Li Y, Liu ET, You M (1993) Risk of miscarriage and a common variant of the estrogen receptor gene. Am J Epidemiol 137: 1361–1364

Thiry-Blaise LM, Taquet A-N, Reginster JY, Nusgens B, Franchimont P, Lapi re CM (1995) Investigation of the relationship between osteoporosis and the collagenase gene by means of polymorphism of the 5'upstream region of this gene. Calcif Tissue Int 56: 88–91

Thomson BM, Saklatvala J, Chambers TJ (1986) Osteoblasts mediate interleukin 1 stimulation of bone resorption by rat osteoclasts. J Exp Med 164: 104–112

Tokita A, Matsumoto H, Morrison NA, Tawa T, Miura Y, Fukamauchi K, Mitsuhashi N, Irimoto M, Yamamori S, Miura M, Watanabe T, Kuwabara Y,

Yabuta K, Eisman JA (1996) Vitamin D receptor alleles, bone mineral density and turnover in premenopausal Japanese women. J Bone Miner Res 11: 1003–1009

Tsai KS, Hsu SH, Cheng WC, Chen CK, Chieng PU, Pan WH (1996) Bone mineral density and bone markers in relation to vitamin D receptor gene polymorphisms in Chinese men and women. Bone 19: 513–518

Uitterlinden AG, Burger H, Huang Q, Grant SFA, Hofman A, vanLeeuwen JPTM, Pols HAP, Ralston SH (1997) Collagen Ia1 Sp1 polymorphism predicts bone mass, bone loss and vertebral fracture. Bone 20: 32S

Uitterlinden AG, Pols HA, Burger H, Huang Q, Daele PLV, Duijn CMV, Hofman A, Birkenhager JC, Leeuwen JPV (1996) A large-scale population-based study of the association of vitamin D receptor gene polymorphism with bone mineral density. J Bone Miner Res 11: 1241–1248

Vandevyver C, Wylin T, Cassiman JJ, Raus J, Geusens P (1997) Influence of the vitamin D receptor gene alleles on bone mineral density in postmenopausal and osteoporotic women. J Bone Miner Res 12: 241–247

Viitanen A, Karkkainen M, Laitinen K, Lamberg-Allardt C, Kainulainen K, Rasanen L, Viikari J, Valimaki MJ, Kontula K (1996) Common polymorphism of the viatmin D receptor gene is associated with variation of peak bone mass in young finns. Calcif Tissue Int 59: 231–234

Woodford-Richens KL, Keen RW, Lanchbury JS, Spector TD (1997) Early postmenopausal bone loss at the spine is associated with polymorphisms at the interleukin 1 receptor antagonist locus but not the interleukin 6 locus. Bone 20: 8S

Yaich L, Dupont WD, Cavener DR, Parl FF (1992) Analysis of the PvuII restriction fragment-length polymorphism and exon structure of the estrogen receptor gene in breast cancer and peripheral blood. Cancer Res 52: 77–83

Zmuda JM, Cauley JA, Danielson ME, Wolf RI, Ferrell RE (1997) Vitamin D receptor gene polymorphisms, bone turnover, and rates of bone loss in older african-american women. J Bone Miner Res 12: 1446–1452

5 The Regulation of Bone Cell Differentiation and Proliferation by Transcription Factors

A.E. Grigoriadis and A. Sunters

5.1 Introduction

One of the basic challenges in bone biology is to identify and understand the mechanisms by which specific molecules regulate cell proliferation, differentiation and cell-cell interactions between bone-forming osteoblasts and bone-resorbing osteoclasts which occur throughout life during normal modeling and remodeling. In recent years, many molecules have been identified which, through very different mechanisms of action, control the balance between formation and resorption. These include for example, cytokines, signaling molecules, receptors, systemic hormones and transcription factors, and these have been roughly mapped to the putative sites of action (Fig. 1). Indeed, each of these molecules are important since alterations in the expression of many of them form the basis of developmental bone defects, metabolic bone

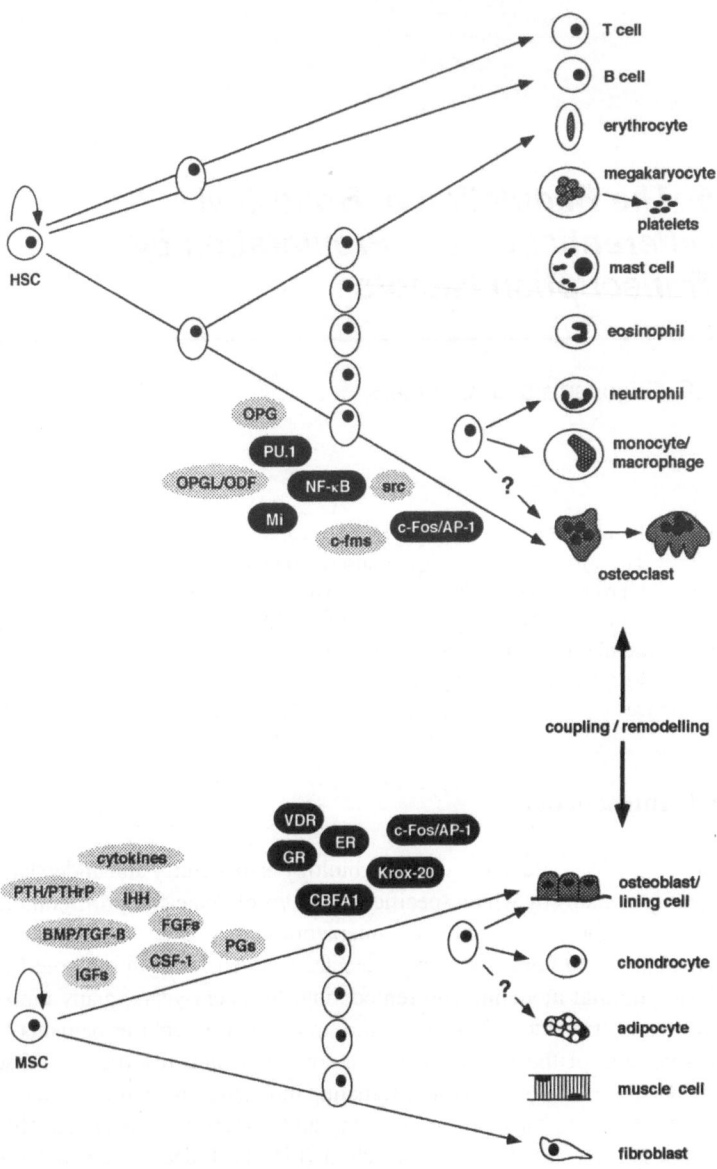

disease and skeletal neoplasias. It can be reasoned that the large number of factors which have been identified to date make it unlikely that a single mechanism alone will explain the molecular basis of bone remodeling. Rather, the intricate control of bone remodeling will likely be due to a combination of factors, some of which share target cells and pathways and others which interact indirectly. Thus, while the large number of factors increase the complexity of this system, they also serve to expand the number of potential targets for intervention.

It is perhaps significant that transcription factors are among this list as it has become evident in recent years that there exist tissue-specific as well as so-called master regulatory genes which have specific effects on bone, thereby giving very important clues as to the mechanisms of bone cell differentiation and activity. More importantly, however, the identification of transcriptional regulators with effects on specific bone cell populations gives the opportunity to isolate novel target genes which act on bone. In this chapter, we will focus on the AP-1 family of transcription factors, focusing specifically on the c-*fos* proto-oncogene and what has been learned about its functional role in bone development and bone disease based on analysis of this molecule in transgenic and knock-out mice.

◄──

Fig. 1. The lineage relationships and the developmental origins of osteoclasts and osteoblasts from hematopoietic stem cells (HSC) and mesenchymal stem cells (MSC), respectively. Also shown are the general lineages which are affected by a variety of molecules that have been demonstrated by different in vitro and in vivo systems to affect bone cell differentiation and coupling between formation and resorption. Molecules are grouped as signaling molecules, including cytokines, peptide hormones, and local growth factors (*gray*) or nuclear transcription factors (*black*). The c-Fos/AP-1 transcription factor is the only one which has been demonstrated in vivo to have essential roles in both hematopoietic (i.e., osteoclasts) and mesenchymal (i.e., osteoblasts) cell lineages. *PTH/PTHrP*, parathyroid hormone/PTH-related protein; *BMP/TGF-β*, bone morphogenetic proteinβ/transforming growth factor-family; *IHH*, indian hedgehog; *FGFs*, fibroblast growth factor family; *IGFs*, insulin-like growth factor family; *CSF-1/c-fms*, colony-stimulating factor-1/c-fms receptor; *PGs*, prostaglandins; *OPG*, osteoprotegerin; *OPGL/ODF-OPG* ligand/osteoclast differentiation factor; *VDR*, vitamin D_3 receptor; *GR*, glucocorticoid receptor; *ER*, estrogen receptor

5.2 The c-*fos* Proto-Oncogene/Transcription Factor: In Vivo Function

c-*fos* is an immediate early gene and is part of a multigene family which includes other *fos*- (*fos*B, *fra*-1, *fra*-2) and *jun*-related (c-*jun*, *jun*B, *jun*D) genes (Angel and Karin 1991). Together, these proteins comprise the AP-1 transcription factor complex which plays an important role in a multitude of processes, including cell proliferation, differentiation, gene expression, apoptosis and oncogenic transformation (Angel and Karin 1991). The complexity of this gene family lies in the multitude of ways it can be regulated: First, as immediate early genes, most *fos*- and *jun*-related genes are tightly regulated at the transcriptional level by hormones and growth factors resulting in short, transient increases in expression. Second, AP-1 activity is dependent upon formation of heterodimers between Fos and Jun proteins and different dimer combinations can result in activation of different target genes. Together with the cell-specific expression of the different Fos and Jun genes, the role of AP-1 will exhibit a high degree of tissue specificity. Third, post-translational modification (e.g. phosphorylation) of these proteins can alter both dimer formation as well as DNA binding and transactivation of target genes. Finally, AP-1 activity can be modified both positively and negatively via protein-protein interactions, for example, with steroid hormone receptors of the glucocorticoid and estrogen receptor families. Faced with this complexity, therefore, the challenge is to identify the tissue-specific function of these molecules in an unambiguous way. To this end, we have analyzed previously c-Fos function in an in vivo context via overexpression in transgenic mice and by gene inactivation in knock-out mice.

To investigate the role of c-Fos during development and oncogenesis, we previously generated transgenic mice harboring a deregulated c-*fos* gene under the control of a ubiquitous promoter element (Grigoriadis et al. 1993). The consequences of high ectopic c-Fos expression were the development of chondroblastic osteosarcomas in virtually all bones of the skeleton with no other apparent phenotype (Fig. 2). Further analysis of this phenotype using in situ hybridization, immunocytochemistry and cell culture techniques demonstrated that the target cells for transformation were cells of the osteoblastic lineage (Grigoriadis et al. 1993). Thus, despite high expression in many transgenic tissues, it appeared that

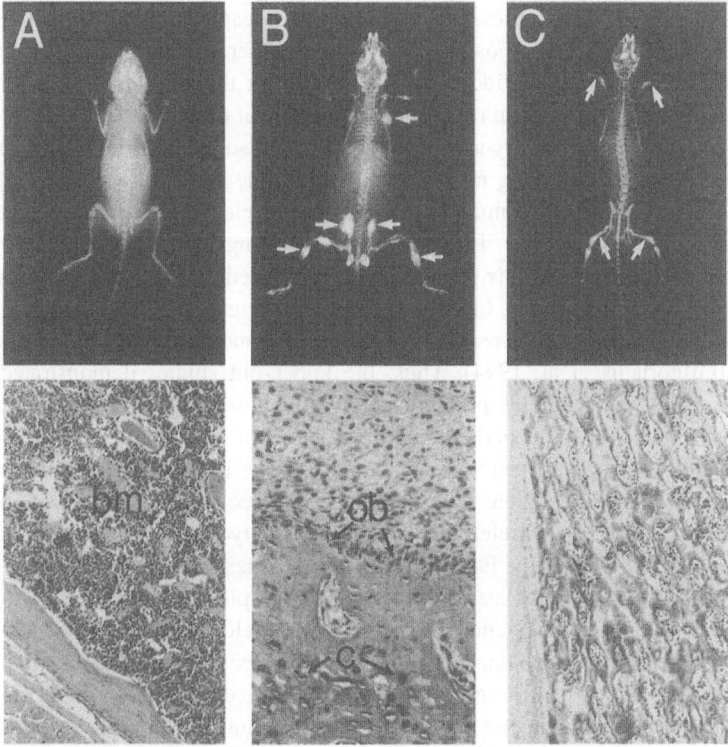

Fig. 2A–C. X-ray (*top panels*) and histological analysis (*bottom panels*) of c-Fos transgenic and knock-out mice. **A** Wild-type mouse showing normal skeletal development and normal long bone architecture containing a bone marrow cavity (*bm*). **B** Overexpression of c-Fos in transgenic mice results in the formation of large calcified tumors (*arrows*) in all bones of the skeleton. Tumors have the histological appearance of chondroblastic osteosarcomas, containing large amounts of neoplastic bone lined by cuboidal osteoblasts (*ob*) and containing some chondrocytes (*c*). **C** Inactivation of c-Fos results in osteopetrosis as seen by the general skeletal sclerosis in the radiograph. Histological analysis shows the absence of a bone marrow space in long bones as it is occupied by unresorbed cartilage and bone (see also Grigoriadis et al. 1995 for further details)

osteoblastic cells represented a unique cell lineage which was sensitive to altered levels of c-Fos protein. These data therefore provided the first indication of a bona fide in vivo target cell for this transcription factor and pointed to a causal role for c-Fos in skeletal neoplasia.

In a complementary approach, we aimed to study the "true" function of c-Fos by analyzing mice in which this gene was inactivated. Quite surprisingly, c-Fos knock-out mice also developed a specific defect associated with bone, i.e., osteopetrosis (Wang et al. 1992). Further characterization of this phenotype demonstrated that the osteopetrosis was due to a complete block in osteoclast differentiation, and there was also a concomitant increase in the number of bone marrow macrophages (Grigoriadis et al. 1994). Thus, the knock-out studies demonstrated unequivocally that the principle in vivo function of c-Fos is osteoclast differentiation, and perhaps also restriction of hematopoietic macrophage-osteoclast precursors. Together with the overexpression studies described earlier, there is compelling evidence that this oncoprotein affects bone tissue preferentially. The specificity of c-Fos for bone was also confirmed by the fact that exogenous expression of other AP-1-related genes in osteoblasts did not develop any phenotype, and inactivation of different *fos*- and *jun*-related genes yielded developmental defects which were distinct from the skeleton (see also Grigoriadis et al. 1995). The c-Fos transgenic and knock-out mice therefore provide excellent tools to further define the downstream pathways and genes which are regulated by c-Fos/AP-1 in specific bone cell populations.

To further confirm the importance of c-Fos and AP-1 in bone tumor formation and to identify cooperating genes, we took a genetic approach: When c-Fos transgenic mice were crossed with c-Jun transgenics, the resulting double transgenic mice developed tumors at a faster rate than single transgenic mice (Fig. 3). Interestingly, the tumors in the Fos/Jun mice appeared more differentiated and more remodeled than in their single transgenic litter mates and this was confirmed by further detailed analysis of the tumors. Fos/Jun tumors as well as cell lines derived from them expressed markedly higher levels of interstitial collagenase, a direct target gene for c-Fos/AP-1, compared to Fos only tumors (Wang et al. 1995). In addition, histochemical staining for tartrate-resistant acid phosphatase (TRAP), an osteoclast marker, indicated a greater number of osteoclasts in Fos/Jun tumors than in Fos tumors (Wang et al. 1995). These results strongly suggested that the presence of

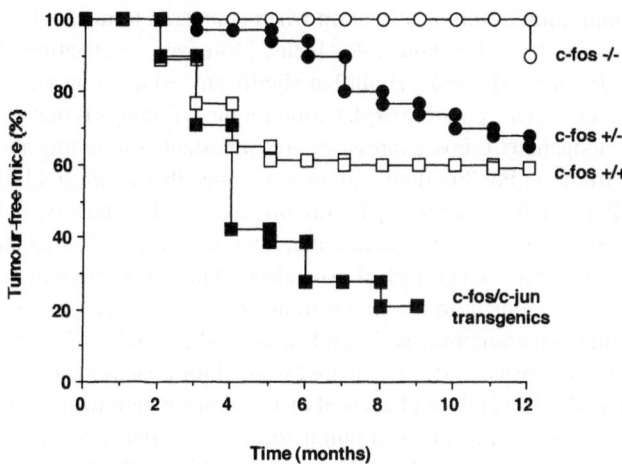

Fig. 3. Frequency of osteosarcoma formation in different compound transgenic mutants overexpressing a c-Fos transgene. Transgenic mice overexpressing both c-Fos and c-Jun oncoproteins (c-*fos*/c-*jun*, *closed squares*) develop tumors faster than littermates overexpressing only c-Fos (c-*fos* +/+, *open squares*). In contrast, the tumors induced by exogenous c-Fos are reduced in mice lacking one endogenous c-*fos* allele (c-*fos* +/–, *closed circles*) and tumors are virtually absent in c-Fos knock-out mice (Fos –/–, *open circles*). All mice were monitored over a period of 9–12 months and the number of mice analyzed in each group varied between 10 (for c-Fos –/–) and 35–55 (for all other genotypes) (see Wang et al. 1995 for additional details)

both Fos and Jun transgenes resulted in the enhancement of two AP-1-dependent events, namely, increased collagenase gene expression and an increase in the number of osteoclasts, both of which are essential for normal bone remodeling as well as for bone tumor progression. Thus, a good correlation was observed between tumor growth and AP-1 gene expression, possibly pointing to a threshold of AP-1 activity which is required for osteoblast transformation and progression to bone tumors.

To further define the role of AP-1, we asked whether decreasing the endogenous c-*fos* levels also affects tumor formation. To this end, Fos transgenic mice were back-crossed onto the Fos knock-out background in which one or both alleles of the endogenous c-*fos* gene were disrupted. Interestingly, Fos null mice expressing a c-Fos transgene showed a marked inhibition of tumor formation when compared to mice which are wild-type at the endogenous c-*fos* locus. Moreover, inactivation of only one c-*fos* allele showed a slight but significant reduction in tumor incidence (Fig. 3). One possible explanation for these findings is that the levels of endogenous c-Fos expression are important, supporting the prediction from the Fos/Jun double transgenic mice that a threshold of Fos and AP-1 activity is necessary for tumorigenesis. Alternatively, and perhaps more intriguingly, is the fact that decreased tumor formation was observed in mice which lacked osteoclasts. Thus, it is tempting to speculate that subsequent to osteoblast transformation, the progression of remodeling osteosarcomas is dependent upon the activity of accessory cells, such as osteoclasts, which are essential for bone remodeling. This is a hypothesis which can be tested in a straightforward manner by investigating whether Fos-induced tumor formation in transgenic mice can be inhibited by in vivo administration of inhibitors of osteoclastic activity, for example, bisphosphonates. Indeed, the use of bisphosphonates to treat the hypercalcemia observed in bone metastases makes this hypothesis feasible and opens the door for identifying novel treatments for primary osteosarcomas.

Taken together, the in vivo analyses of c-Fos function in mice provided an unequivocal link between c-Fos/AP-1 and bone tumor formation. Interestingly, high c-Fos expression has been documented in the majority of primary and metastatic human osteosarcomas, as well as in bone cells of Paget's disease patients, a proportion of which develop secondary osteosarcoma (Wu et al. 1990; Hoyland and Sharpe 1994; Pompetti et al. 1996). Primary human tumors and osteosarcoma cell lines also frequently show mutations and/or deletions in the p53 and pRB tumor suppressor genes (Masuda et al. 1987; Nigro et al. 1989) and, most recently, alterations in some cell cycle genes such as cyclin D1, CDK4 and p16 (see below) have been observed in osteosarcoma (Maelandsmo et al. 1995). Although these studies are largely descriptive, our in vivo observations together with the data from human osteosarcomas suggest a causal role of c-Fos in bone tumor formation and imply that interference with the cell cycle machinery is likely to play an important role.

5.3 c-Fos and Bone: Is There a Link to the Cell Cycle?

It is now quite well established that tissues which show a dependency on specific genes to control their proliferation and differentiation may be driven to malignancy by alterations in those same genes. In this regard, c-Fos provides an ideal example of a transcription factor whose endogenous function is essential for normal bone development, but whose deregulated expression results in bone tumor formation. Moreover, alterations in the basic machinery of cell proliferation, namely, the cell cycle, have important functional consequences for normal development as well as in different types of neoplasias. The eukaryotic cell cycle is regulated by a variety of proteins which interact to ensure that inappropriate cell division does not occur and that all the phases of the cycle occur in the correct chronological order (Harper and Elledge 1996). The main regulatory proteins involved are cyclins, cyclin-dependent kinases (CDKs) and CDK inhibitors (CKIs) (Fig. 4). Cyclin expression typically oscillates throughout the cell cycle, such that cyclins D and E are expressed in the G1 phase, whereas cyclins A and B are commonly associated with M and S phases (Hunter and Pines 1994). Cyclins function by forming complexes with specific CDKs, which become activated by phosphorylation (Hunter and Pines 1994). Thus, in G1, E- and D-type cyclins associate with CDK2 and CDK4 or CDK6 respectively, resulting in phosphorylation of the retinoblastoma gene product (pRB). Phosphorylated pRB dissociates from the transcription factor E2F, allowing E2F to stimulate the transcription of growth-related genes and allow progression through G1 and S phases. Exit from the cell cycle and maintenance of cells in the nonproliferative state are carried out by CKIs, which bind to and inhibit the activity of different cyclin/CDK complexes. There exist two classes of CKIs which differ in their targets of inhibition. The CIP/KIP family inhibits a broad range of kinases and consists of p21, p27 and p57 proteins, whereas the INK4 family members, containing p15, p16, p18 and p19, specifically inhibit CDK4 and CDK6 (Sherr and Roberts 1995; Elledge et al. 1996).

With respect to oncogenesis, the pRB and p53 tumor suppressor genes are often mutated, inactivated or deleted in many human and mouse tumors, including a high proportion of osteosarcomas (Weinberg 1995). Overexpression of cyclin D1 in vitro as well as in vivo in transgenic mice has demonstrated that cyclin D1 behaves like an onco-

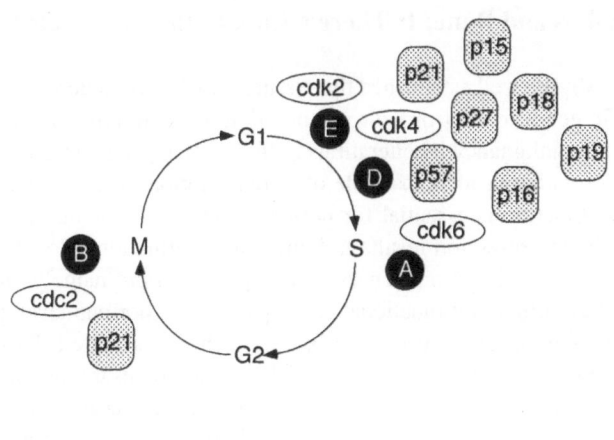

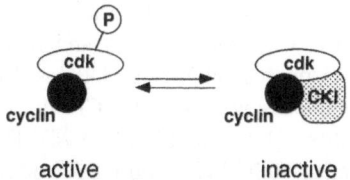

active   inactive

Fig. 4. The basic components of the cell cycle machinery and the approximate phase of the cell cycle in which they have been shown to act. The *bottom panel* indicates the basic stoichiometry of the ternary complex which is formed between cyclins (*black*), cyclin-dependent kinases (cdk; *white*) and kinase inhibitors (CKI; *shaded*). Complexes are activated by specific phosphorylation (*P*) and inhibited by binding to specific CKIs

gene and is overexpressed in many primary carcinomas (Hunter and Pines 1994) as well as in parathyroid adenomas, which lead to increased bone resorption and hypercalcemia (Motokura et al. 1991). The finding, that some CKIs (e.g., p16) act as tumor suppressors, provides a direct link between tumorigenesis and disruption of cell cycle control, and there are an increasing number of examples of knock-out mice lacking different CKIs (p21, p27, p16, p57) which develop specific hyperplasias and tumors in the absence of CKI function (Elledge et al. 1996).

While aberrations in normal cell cycle control have been tightly associated with tumorigenesis and oncogene function, it is not well understood whether there is a functional link between c-Fos and the cell cycle. In vivo, a recent report using an inducible c-Fos construct in transgenic mice has implicated c-Fos in p27 stability as c-Fos induction inhibited p27 down-regulation following B cell activation (Kobayashi et al. 1997). Antisense and antibody microinjection experiments in vitro as well as conditional transformation systems in fibroblasts have suggested variable roles for c-Fos in cell cycle progression, although the growth of embryonic fibroblasts isolated from c-Fos knock-out mice is apparently unaffected (Brüsselbach et al. 1995). Connections between AP-1 family members and cyclin D1 have been shown, as c-Fos can both increase growth rate and cyclin D1 transcription, as well as decrease growth rate and inhibit cyclin D1, cdc2 and CDK4 transcription (Miao and Curran 1994; Balsabore and Jolicoeur 1995). These results suggest that c-Fos displays different effects in particular fibroblastic cell lines in vitro. With respect to bone, the study of the cell cycle in osteoblast proliferation and differentiation has not been investigated extensively. A recent study has shown that cyclins E and B are up-regulated during osteoblast differentiation in vitro but the activity of cyclin/CDK complexes and the expression of CKIs was not evaluated (Smith et al. 1995).

Thus, the link between AP-1 activity and cell cycle control in normal and malignant osteoblasts is unclear at this time. With a suitable in vivo model in which c-Fos affects osteoblast proliferation in transgenic mice, we were able to ask whether normal cell cycle gene expression in osteoblasts is perturbed in the presence of deregulated c-Fos activity. The potential findings will therefore elucidate the possible mechanisms underlying osteoblast growth control and therefore have significant implications for identifying novel modulators of osteoblast proliferation.

5.4 Cell Cycle Gene Expression in Osteoblasts

Initial studies focused on investigating the expression patterns of several cell cycle-associated genes by in situ hybridization and immunocytochemistry in normal osteoblasts and in c-Fos-transformed osteoblasts within bone tumors. Of the cyclins, cyclin E was expressed in embry-

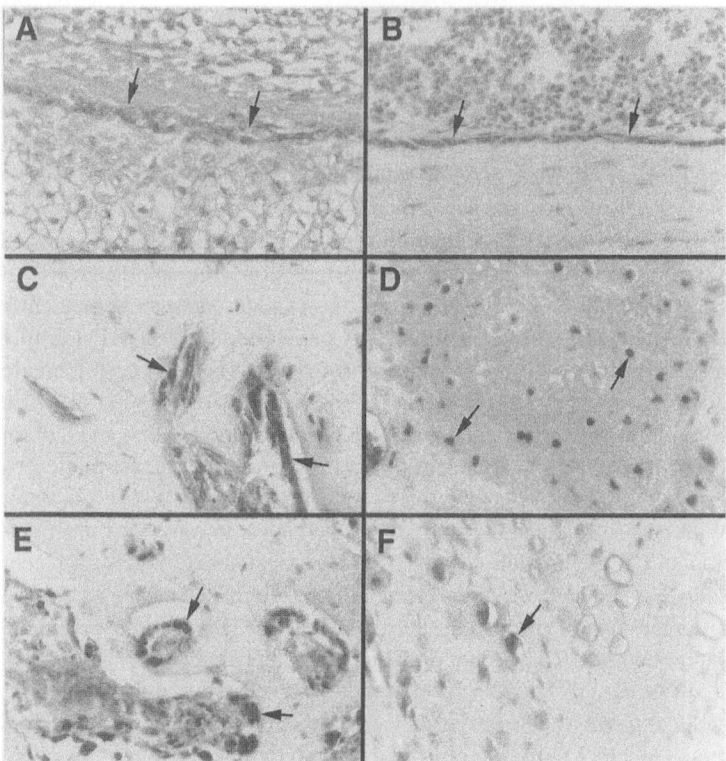

Fig. 5A–F. Immunocytochemical staining of different cell cycle-associated proteins in developing mouse bone as well in c-Fos-induced osteosarcomas in transgenic mice. **A** Expression of cyclin E in osteoblasts present in the developing bony collar of a 16.5 day fetal wild-type metatarsal. **B** Expression of cyclin E is restricted to endosteal osteoblasts in a 20 day-old post-natal tibia. Cyclin D1 is expressed at very high levels in osteoblasts (**C**) and chondrocytes (**D**) present within transgenic chondroblastic osteosarcomas. **E** Cyclin-dependent kinase (CDK)4 is also expressed at high levels in transformed osteoblasts. **F** High levels of the cyclin kinase inhibitor (CKI) p27 in some chondrocytes within the tumors. All sections are 5 m paraffin sections, stained with polyclonal antibodies to each specific cell cycle component and counterstained with hematoxylin

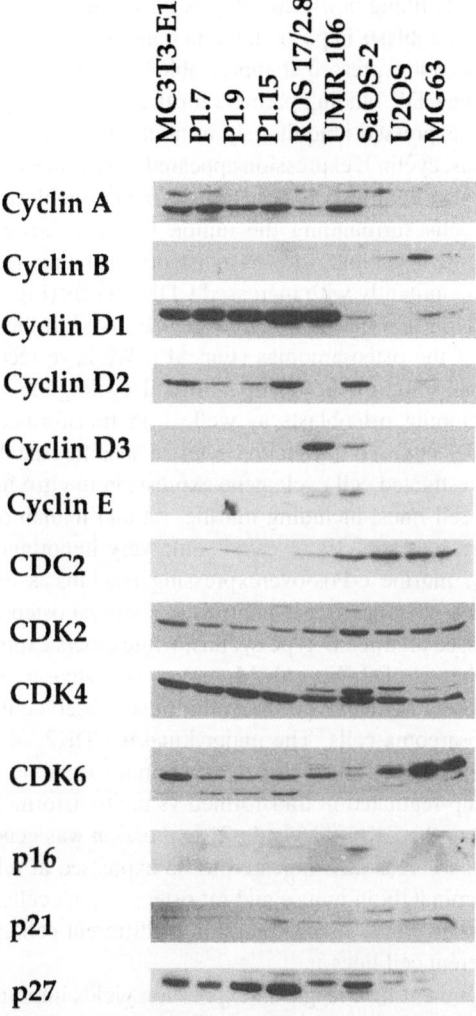

Fig. 6. Western blot analysis of cell cycle gene expression in a variety of non-transformed osteoblasts and transformed osteosarcoma cells. Murine cell lines include MC3T3-E1, and P1.7, P1.9, P1.15 cells which were derived from c-Fos transgenic osteosarcomas (see Grigoriadis et al. 1993). Rat osteosarcoma cell lines were ROS 17/2.8 and UMR 106, and human osteosarcoma cells were SaOS-2, U2OS and MG63. Protein extracts and SDS-PAGE analysis were performed following standard protocols and Western blotting was carried out using specific antibodies to each cell cycle component

onic osteoblasts lining newly forming bone surfaces and was restricted to endosteal osteoblasts in postnatal long bones (Fig. 5A,B). In contrast, cyclin D1 was not expressed at appreciable levels in normal osteoblasts in situ, and the CDK binding partners for these cyclins, e.g., CDK2 and CDK4, were expressed ubiquitously in nontransformed osteoblasts. In osteosarcomas, cyclin E expression appeared to be more widespread and was expressed ubiquitously in both transformed osteoblasts as well as in fibroblastic cells surrounding the tumor. Of note, however, was the marked increase in cyclin D1 levels in osteoblasts lining neoplastic bone surfaces, concomitantly with increased CDK4 levels (Fig. 5C–E). With respect to CKIs, high levels of p27 were observed in the chondrocyte component of the osteosarcomas (Fig. 5F). We have recently characterized further some of the changes in cell cycle gene expression in normal developing osteoblasts as well as in transformed osteoblasts (Sunters et al. 1998). To further characterize specific osteoblast populations, we investigated cell cycle gene expression in vitro in a variety of osteoblastic cell lines, including murine, rat and human osteosarcoma cells. Western blot analysis revealed some very important differences: Significantly, murine c-Fos-overexpressing osteoblasts expressed cyclins D1 and D2 but not D3, in contrast to some rat osteosarcoma cells which expressed all three D-type cyclins, while others expressed cyclins D1 and D3 but not D2 (Fig. 6). Moreover, rat osteosarcoma cells expressed markedly higher levels of cyclin E than observed in murine and human osteosarcoma cells. The major kinases CDK2, -4 and -6 were expressed uniformly in all cell types with minor variations, while CDC2 appeared to up-regulated in transformed vs nontransformed bone cells. With respect to the CKIs, p16 and p21 expression was generally low in all cell lines, whereas p27 appeared to be expressed at relatively high levels, predominantly in mouse and rat osteosarcoma cells. Thus, there appears to be heterogeneous expression of different cell cycle components in different cell lines.

While documenting changes in expression yields important information, the signals which determine cell cycle progression depend ultimately upon the activity of the ternary complex formed between cyclins, CDKs and CKIs (Fig. 4). Indeed, changes in complex composition and activity have been documented in the absence of changes in RNA or protein levels. Thus, studies which determine the composition and activity of specific complexes in the presence and absence of c-Fos will be more informative. Co-immunoprecipitations in rat osteosarcoma cells have determined that cyclin E/CDK2 dimers are bound to p27 (data not shown), confirming the high expression of this CKI seen in these cells. Clearly, the next step is to perform kinase assays to determine the functionality of these and similar complexes in the different osteoblastic cell lines.

Thus, these data suggest that different genes may control cell cycle progression in specific osteoblastic cell types. The future challenge therefore is to identify the basis of this heterogeneity and investigate whether different osteoblastic populations are regulated differently. Since different osteoblastic cell populations (e.g., trabecular osteoblasts, periosteal vs endosteal osteoblasts, lining cells) are preferentially affected in different metabolic bone diseases, identification of specific cell cycle gene expression and complex activity in these populations is necessary to understand the growth requirements of these cells.

5.5 Studies Using Inducible c-Fos Systems

Finally, in order to define unequivocally the role of c-Fos in cell cycle regulation in osteoblasts, and to identify direct targets for this transcription factor which control osteoblast proliferation and activity, it would be highly desirable to be able to activate c-Fos expression at will in different cell populations. To this end, we have generated an inducible system in which c-Fos expression is tightly regulated by tetracycline and its analogues (Gossen et al. 1995; Lang and Feingold 1996). Several clones of MC3T3-E1 osteoblastic cells have been generated and Northern blot analysis has confirmed that the c-Fos transgene can be induced in a tetracycline-dependent manner (Fig. 7). With respect to cell cycle genes, preliminary experiments using these cell lines have indicated that the expression of c-Fos causes the induction of cyclin D1 and p27 expression and inhibition of cyclin D2 expression (data not shown). These preliminary results are very exciting and suggest that specific cell cycle genes can be differentially regulated by ectopic c-Fos expression, at least in MC3T3-E1 osteoblastic cells. The functional consequences of this differential gene regulation with respect to cell cycle progression and transformation remains to be elucidated. Finally, the strengths of this inducible system lie in the potential to identify novel regulators of

Fig. 7. Expression of c-*fos* mRNA in two MC3T3-E1 clones (pJMF2-Fos 2.1 and pJMF-Fos 2.6) harboring a tetracycline-regulated c-*fos* vector. Expression of the c-*fos* transgene is suppressed in the presence of tetracycline and is activated following tetracycline withdrawal. The increase in mRNA expression is paralleled by increases in c-Fos protein levels as confirmed by Western blot analysis (data not shown)

osteoblast proliferation which are stimulated by the c-Fos transcription factor.

5.6 Conclusions and Perspectives

It is quite well accepted that much of what we know about the biology of bone cell proliferation, differentiation and bone remodeling comes from investigating different metabolic bone diseases and skeletal neoplasias. Thus, having a model system in which different parameters of osteoblast activity can be investigated in a highly reproducible manner in an in vivo context would be highly advantageous. The mutant mice generated for analyzing the function of the c-Fos proto-oncogene and transcription factor provide one such useful model. Unlike molecules such as peptide hormones, growth factors and receptor antagonists, it may be perhaps difficult to envision nuclear transcription factors as providing immediate targets for developing compounds towards therapy of bone disorders such as osteoporosis. However, the strengths of research into transcription factor biology are essentially two-fold: First, by elucidating mechanisms which lie downstream of transcription factor activation, there is a unique opportunity to identify novel osteoblast-specific genes which regulate different aspects of the osteoblast phenotype and which can provide novel targets for therapeutic intervention. Secondly, one of the most exciting aspects of these molecules is not only their ability to directly activate target gene transcription, rather their ability to interact with other transcription factors via protein-protein interaction and therefore modulate transcriptional activity. For AP-1, interactions with glucocorticoid and estrogen receptor complexes have far reaching implications for bone. Indeed, the recent report describing the transactivation properties of the two estrogen receptors (ERα and ERβ) at AP-1 sites (Paech et al. 1997) opens the door for investigating ER-AP-1 interactions in osteoblasts in combination with the continuing progress with synthetic selective estrogen receptor modulators. Clearly, the tools provided by the c-Fos transgenic and knock-out mice are ideal for addressing some of these issues in a well-defined way. The tumor-bearing mice will not only shed light on the mechanisms of oncogenesis, but they will identify the major players and novel genes which are important for the tight growth control of normal osteoblasts. Together with the ability to assess gene function using the current mouse technology (see Chap. 6), the analysis of transcription factors in bone can provide a whole array of previously undescribed or unidentified molecules which would be a great opportunity for developing novel pharmaceuticals.

Acknowledgements. The c-Fos transgenic and knock-out mice were generated in the laboratory of Dr. E.F. Wagner (IMP, Vienna).

References

Angel P, Karin M (1991) The role of Jun, Fos and the AP-1 complex in cell-proliferation and transformation. Biochim Biophys Acta 1072:129–157

Balsabore A, Jolicoeur P (1995) Fos proteins can act as negative regulators of cell growth independently of the fos transforming pathway. Oncogene 11:455–465

Brüsselbach S, Möhle-Steinlein U, Wang Z-Q, Schreiber M, Lucibello FC, Müller R, Wagner EF (1995) Cell proliferation and cell cycle progression are not impaired in fibroblasts and ES cells lacking c-Fos. Oncogene 10:79–86

Elledge SJ, Winston J, Harper JW (1996) A question of balance – the role of cyclin-kinase inhibitors in development and tumorigenesis. Trends Cell Biol 6:388–392

Gossen M, Freundlieb S, Bender G, Muller G, Hillen W, Bujard H (1995) Transcriptional activation by tetracyclines in mammalian cells. Science 268:1766–1769

Grigoriadis AE, Schellander K, Wang Z-Q, Wagner EF (1993) Osteoblasts are target cells for transformation in c-*fos* transgenic mice. J Cell Biol 122:685–701

Grigoriadis AE, Wang Z-Q, Cecchini MG, Hofstetter W, Felix R, Fleisch HA, Wagner EF (1994) c-Fos is a key regulator of osteoclast/macrophage lineage determination and bone remodeling. Science 266:443–448

Grigoriadis AE, Wang Z-Q, Wagner EF (1995) *Fos* and bone cell development: lessons from a nuclear oncogene. Trends Genet 11:436–441

Harper JW, Elledge SJ (1996) CDK inhibitors in development and cancer. Curr Opin Genet Devel 6:56–64

Hoyland J, Sharpe PT (1994) Up-regulation of c-Fos protooncogene expression in pagetic osteoclasts. J Bone Mineral Res 9:1191–1194

Hunter T, Pines J (1994) Cyclins and cancer. II. Cyclin-D and CDK inhibitors come of age. Cell 79:573–582

Kobayashi K, Phuchareon J, Inada K, Tomita Y, Koizumi T, Hatano M, Miyatake S, Tokuhisa T (1997) Overexpression of c-fos inhibits down-regulation of a cyclin-dependent kinase-2 inhibitor p27(Kip1) in splenic B cells activated by surface Ig cross-linking. J Immunol 158:2050–2056

Lang Z, Feingold JM (1996) An autonomously replicating eukaryotic expression vector with a tetracycline-responsive promoter. Gene 168:169–171

Maelandsmo GM, Berner J-M, Florenes VA, Forus A, Hovig E, Fodstad O, Myklebost O (1995) Homozygous deletion frequency and expression levels of the CDKN2 gene in human sarcomas – relationship to amplification and messenger RNA levels of CDK4 and CCND1. Brit J Cancer 72:393–398

Masuda H, Miller C, Koeffler HP, Battiflora H, Cline MJ (1987) Rearrangement of the p53 gene in human osteogenic sarcomas. Proc Natl Acad Sci USA 84:7716–7719

Miao GG, Curran T (1994) Cell transformation by c-Fos requires an extended period of expression and is independent of the cell cycle. Mol Cell Biol 14:4295–4310

Motokura T, Bloom T, Kim HG, Jüppner H, Ruderman JV, Kronenberg HM, Arnold A (1991) A novel cyclin encoded by a BCL1-linked candidate oncogene. Nature 350:512–515

Nigro JM, Baker SJ, Preisinger AC, Jessup JM, Hostetter R, Clearly K, Bigner SH, Davidson N, Baylin S, Devilee P, Glover T, Collins FS, Weston A, Modali R, Harris CC, Vogelstein B (1989) Mutations in the p53 gene occur in diverse human tumor types. Nature 342:705–708

Paech K, Webb P, Kuiper GGJM, Nilsson S, Gustafsson J-Å, Kushner P, Scanlan TS (1997) Differential ligand activation of estrogen receptors ERα and ERβ at AP1 sites. Science 277:1508–1510

Pompetti F, Rizzo P, Simon RM, Freidlin B, Mew DJ, Pass HI, Picci P, Levine AS, Carbone M (1996) Oncogene alterations in primary, recurrent and metastatic human bone tumors. J Cell Biochem 63:37–50

Sherr CJ, Roberts JM (1995) Inhibitors of mammalian G(1) cyclin-dependent kinases. Genes Dev 9:1149–1163

Smith E, Frenkel B, Schlegel R, Giordano A, Lian JB, Stein LL, Stein GS (1995) Expression of cell cycle regulatory factors in differentiating osteoblasts – postproliferative up-regulation of cyclin-B and cyclin-E. Cancer Res 55:5019–5024

Sunters A, McCluskey J, Grigoriadis AE (1998) Control of cell cycle gene expression in bone development and during c-Fos-induced osteosarcoma formation. Devel Genet 22:386–397

Wang Z-Q, Liang J, Schellander K, Wagner EF, Grigoriadis AE (1995) c-Fos-induced osteosarcoma formation in transgenic mice: Cooperativity with c-Jun and the role of endogenous c-Fos. Cancer Res 55:6244–6251

Wang Z-Q, Ovitt C, Grigoriadis AE, Möhle-Steinlein U, Rüther U, Wagner, EF (1992) Bone and haematopoietic defects in mice lacking c-fos. Nature 360:741–745

Weinberg RA (1995) The retinoblastoma protein and cell cycle control. Cell 81:323–330

Wu J-X, Carpenter PM, Gresens C, Keh R, Niman H, Morris JWS, Mercola D (1990) The proto-oncogene c-Fos is overexpressed in the majority of human osteosarcomas. Oncogene 5:989–1000

6 Transgenic and Gene Knock-Out Animals in Skeletal Research

M. Amling, M.W. Hentz, M. Priemel, and G. Delling

6.1 Introduction

Until recently most of our knowledge about the skeleton was derived from descriptive morphology, histomorphometry, endocrinology, and cellular studies of bone turnover (Delling 1987; Delling and Amling 1995; Amling et al. 1994, 1996). Recent approaches have led to the identification of local factors that regulate skeletal morphogenesis. Molecular and biochemical studies of bone cells in vitro and, most importantly, the power of genetics entering the bone field have led us toward the beginning of a molecular understanding of the skeletal system (Erlebacher et al. 1995). Indeed the identification of genes responsible for mouse and human skeletal abnormalities, gene inactivation and targeted gene misexpression in mice have documented the importance of specific signaling molecules and their receptors, as well as growth factors, matrix proteins, and transcription factors in the development and maintenance of bone. The successful convergence of mouse and human genetics in skeletal biology has been demonstrated several times, e.g. chondrodysplasia in parathyroid hormone-related protein (PTHrP) receptor mutant mice (Amizuka et al. 1994; Karaplis et al. 1994; Lanske et al. 1996; Weir et al. 1996; Amling et al. 1997) and patients with Jansens metaphyseal dysplasia (Schipani et al. 1995). Mutations in collagen type XI in *cho/cho* mice (Li et al. 1995) and patients with Stickler syndrome (Vikkula et al. 1995); identical phenotypes of *Cbfa1*+/– heterozygous mice (Otto et al. 1997) and patients with cleidocranial dysplasia (Mundlos et al. 1997), which lack the expression of one allele of the *Cbfa1* gene; as well as mice with targeted ablation of the second zinc finger of the vitamin D receptor DNA-binding domain as a model of vitamin D-dependent rickets type II (Li et al. 1997). The latter examples underscore the invaluable importance of transgenic and knock-out animals in skeletal research.

One has to keep in mind, however, that so far the results from mouse models have been mixed. Take the mouse in which the retinoblastoma tumor-suppressor gene (*Rb*) was knocked out. In humans, the lack of Rb leads to a cancer in the retina of the eye. But when the gene is inactivated in mice, the animals get pituitary gland tumors. Furthermore, mice lacking *BRCA1*, which were supposed to be a model for human breast and ovarian cancer, do not develop any tumors at all. Thus, it seems that in mice, other genes can compensate for a missing gene, such as *BRCA1*, and that the genetic wiring for growth control in mice and human is subtly different.

Therefore to be an important tool for furthering our understanding of the skeleton, transgenic and knock-out animals need to be designed and analyzed in view of bone biology and physiology. This background defines the potential targets for specific mutation of defined genes in the germ line of mice. Future progress in skeletal research by transgenic animals will thus markedly depend on: (1) a bone specific aspect, in the form of choosing a promising target gene, cell, function, and stage, and (2) a general aspect, using a genetic approach.

6.2 Bone Biology Defines Targets in Skeletal Research

The development and maintenance of the skeleton is ultimately the result of coordinated cellular differentiation, function, and interaction. Four major cell types contribute to the skeleton: (1) chondrocytes, which form cartilage; (2) osteoclasts, the only cell capable of resorbing bone; (3) osteoblasts, which are responsible for bone formation; and (4) osteocytes, representing terminally differentiated osteoblasts embedded in bone matrix and thought to be involved in mechanotransduction.

In general, two types of bone formation are distinguished: intramembranous and endochondral bone formation. During intramembranous bone formation mesenchymal precursor cells directly differentiate into bone forming osteoblasts. Examples are the formation of flat bones of the skull, the formation of part of the clavicle, and the bone apposition to the periosteal surface of diaphyseal cortex of long bones. In contrast, endochondral bone formation represents the conversion of an initial cartilage template into bone and is responsible for generating the majority of bones. During endochondral ossification, the chondrocytes present in the early cartilaginous model, and later in the growth plate, first proliferate and then progressively differentiate into mature hypertrophic chondrocytes. Once fully differentiated, these hypertrophic cells participate in the mineralization of the cartilaginous matrix and undergo cell death. In normal bone development, this is followed by – and may be the necessary signal for – the local recruitment of blood vessels and osteoclasts into the zone of provisional mineralization, leading to the progressive replacement of cartilage by bone, the homing of the hematopoietic

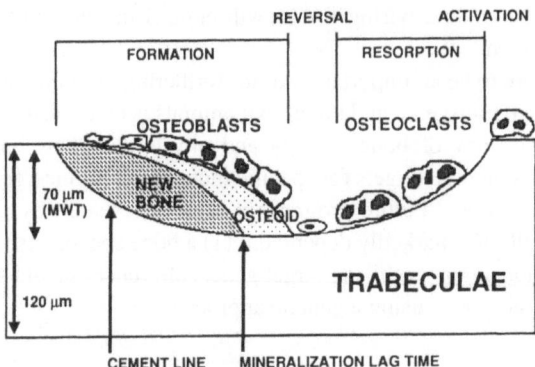

Fig. 1. Remodeling scheme. In physiological situations bone resorption and bone formation are balanced to maintain a stable bone mass. The mechanisms which control this cellular interaction remain, however, poorly understood

bone marrow and ultimately longitudinal bone growth. Throughout life, vertebrates constantly renew bone through remodeling, which is characterized by two successive phases: resorption of existing bone by osteoclasts followed by new bone formation by the osteoblasts. Under physiological conditions, bone resorption and bone formation are balanced to maintain bone structure and bone volume (Fig. 1). Moreover, the cellular interaction of osteoclasts and osteoblasts connects the skeleton as the major reservoir of calcium to the endocrine regulation of ion homeostasis in humans.

Thus, potential targets in skeletal research can be divided into two groups: (1) those involving either a specific cell type or genes which are specifically expressed in one bone cell type, or a special function, either bone formation or bone resorption, and (2) those involving a specific stage during life, either skeletal development or skeletal maintenance.

6.3 Skeletal Development, Chondrocyte Differentiation, and Bone Growth

The two major milestones in the recent understanding of skeletal development are the significant molecular insights into the differentiation of the osteogenic lineage and the progress in understanding endochondral bone formation.

6.3.1 Osf2/Cbfa1:
A Master Key to Osteoblastic Bone Formation

A series of four papers published in the journal *Cell* this summer provided compelling evidence that Osf2/Cbfa1 is an essential transcription factor required for osteoblast differentiation and thus the master key to bone formation. *Cbfa* belongs to the Runt domain gene family, homologous to the *Drosophila melanogaster* pair-rule gene *runt* that plays a role in the formation of the segmented body pattern as well as in the development of the nervous system and in sex determination (Ogawa et al. 1993). In two independent studies Komori et al. (1997) and Otto et al. (1997) deleted *Cbfa1* and observed the complete lack of bone formation in *Cbfa1*-deficient mice, while *Cbfa1*+/– heterozygous mice presented a phenotype paralleling the findings in patients with cleidocranial dysplasia (CCD) (Jensen 1990). Knowing the findings of Otto et al (1997) in mice, Mundlos et al. (1997) were able to demonstrate the lack of expression of one allele of the *Cbfa1* gene in patients with CCD. Finally Karsentys group provided direct evidence that Cbfa1 acts as a transcription factor that up-regulates the expression of osteoblast-related genes (Ducy et al. 1997). Most recently we have developed evidence that *Osf2/Cbfa1* controls its own expression and is required for postnatal bone formation (Ducy, Amling, Karsenty, unpublished data). This demonstrates that Osf2 has a dual function. Besides its role during development, it is a positive regulator of bone formation by pre-existing osteoblasts after birth. For a more detailed review of the regulation of osteoblast differentiation and the role of transcription factors in bone cells, the reader is referred to chapters in this volume by Beresford and Grigoriadis.

6.3.2 Endochondral Bone Formation

Although the different genetic models of endochondral bone formation result in a huge variety of skeletal phenotypes, their unifying principle is the disturbed timing of chondrocyte differentiation: A delay of chondrocyte differentiation is observed in humans with Jansens metaphyseal dysplasia due to constitutive activation of the PTH/PTHrP receptor (Schipani et al. 1995) as well as in mice after targeted overexpression of PTHrP to chondrocytes (Weir et al. 1996), overexpression of fibroblast growth factor-2 (FGF-2) (Coffin et al. 1995), and targeted deletion of insulin-like growth factor-1 (IGF-1) (Baker et al. 1993) causing a phenotype paralleling that of the growth hormone (GH)-deficient Snell dwarf mice (Snell et al. 1929). By contrast, deletion of the gene coding for PTHrP (Amizuka et al. 1994; Karaplis et al. 1994), or its receptor (Lanske et al. 1996), as well as of the *bcl*-2 gene (Amling et al. 1997; Veis et al. 1993; Nakayama et al. 1994) results in an acceleration of this developmental program with premature terminal chondrocyte differentiation. Other important factors in growth plate control are the FGF-3 receptor and activating transcription factor-2 (ATF-2). A gain of function mutation in the FGF-3 receptor results in achondroplasia due to decreased chondrocyte proliferation (Shiang et al. 1994; Rousseau et al. 1994, 1995), as does the lack of activating transcription factor-2 (ATF-2) (Reimhold et al. 1996), while the FGF-3 receptor knock-out mice show a growth plate with an increased proliferating and hypertrophic zone (Deng et al. 1996; Muenke and Schell 1995).

6.3.3 Indian Hedgehog-Parathyroid Hormone-Related Peptide-Bcl-2: A Signaling Pathway in the Control of Endochondral Ossification

Parathyroid hormone-related peptide (PTHrP) was first isolated from human carcinomas (Strewler et al. 1987; Suva et al. 1987; Mangin et al. 1988) and is the causative agent for the humoral hypercalcemia associated with various malignancies. PTHrP is structurally related to PTH, a hormone of major importance in calcium metabolism (Kronenberg et al. 1993). Both peptides share 8 of 13 amino-terminal residues, and bind to and activate the same G-protein-coupled PTH/PTHrP receptor (Jüppner et al. 1988, 1991). Unlike PTH, however, PTHrP does not circulate in appreciable amounts in normal subjects but is instead widely expressed in fetal and adult tissues, where it is thought to regulate cell differentiation, cell proliferation and organogenesis as a paracrine or autocrine soluble factor (Goltzman et al. 1989; Wysolmerski et al. 1994; Broadus and Stewart 1994). In this context, PTHrP is a mediator of cellular growth and differentiation (Karaplis et al. 1994; Amizuka et al. 1994) and is involved in mesenchymal-epithelial interactions in several tissues (Hardy 1992; van de Stolpe et al. 1993; Wysolmerski et al. 1994).

The critical role played by PTHrP and its receptor in skeletal development has recently been demonstrated unequivocally by gene targeting and disruption experiments in mice and a natural mutation in humans. Mice homozygous for ablation of the PTHrP gene or the PTH/PTHrP receptor gene die at birth and exhibit skeletal deformities that are due, at least in part, to a decrease in proliferation and the accelerated differentiation of chondrocytes in the developing skeleton (Karaplis et al. 1994; Amizuka et al. 1994, 1996a, 1996b; Lanske et al. 1996). The endochondral bones of these animals are shorter, wider, deformed, and undergo premature mineralization. At the other end of the spectrum, striking skeletal deformities are observed in Jansens metaphyseal chondrodysplasia. This rare human genetic disorder characterized by short-limbed dwarfism with delayed endochondral maturation and agonist-independent hypercalcemia that has been attributed to an activating PTH/PTHrP receptor mutation that results in constitutive cAMP accumulation (Jansen 1934; Schipani et al. 1995; Jüppner 1996).

The patterning gene Indian hedgehog (Ihh) regulates the expression of the central signaling molecule PTHrP (Vortkamp et al. 1996; Lanske et al. 1996), which activates its receptor and exerts, at least in part, its control of chondrocyte maturation by stimulating expression of Bcl-2, a protein that controls programmed cell death in several cell types, and delays the terminal differentiation and apoptosis in chondrocytes (Allsopp et al. 1993; Boise et al. 1993; Korsmeyer 1992; Hockenbery et al. 1993; Oltvai et al. 1993; Amling et al. 1997) (Fig. 2).

Consequently targeted overexpression of PTHrP in chondrocytes using the mouse collagen II promoter was found to result in overexpression of Bcl-2 and a striking form of chondrodysplasia characterized by an accumulation of chondrocytes in their prehypertrophic stage and

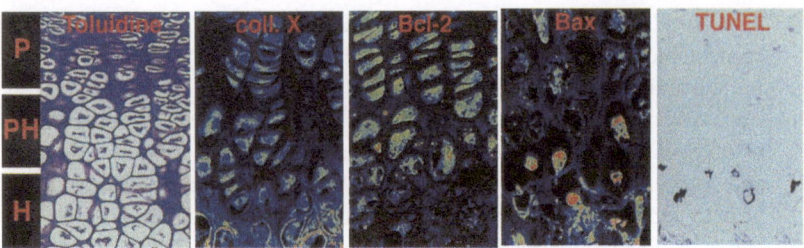

Fig. 2. Legend see p. 131

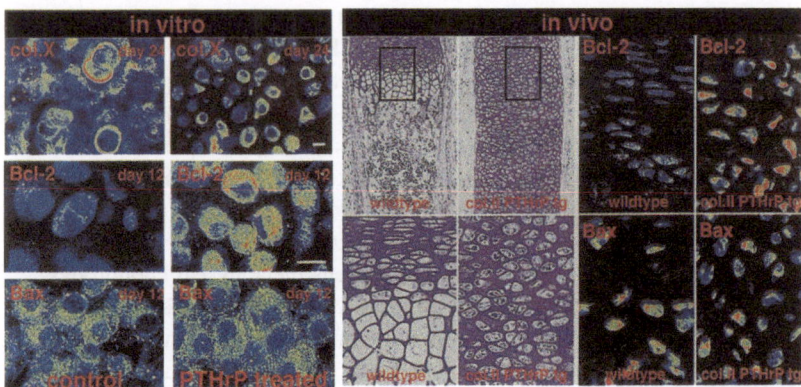

Fig. 3. Legend see p. 131

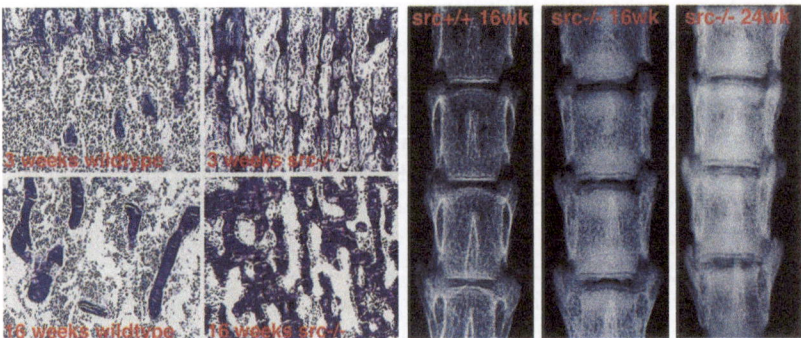

Fig. 10. Legend see p. 131

severely delayed endochondral bone formation (Weir et al. 1996) (Fig. 3).

Bcl-2 is indeed directly involved, and not just a bystander protein, in endochondral bone formation as demonstrated by premature maturation of chondrocytes in *bcl*-2 knock-out mice, where Bcl-2 levels are manipulated independently of PTHrP or any other molecule (Fig. 4).

These observations have led to a new model for the control of chondrocyte differentiation (Fig. 5). Preliminary data analyzing the co-expression of PTHrP and Bcl-2 in human chondrosarcomas confirm further the importance of the PTHrP/Bcl-2 pathway, at least in chondrogenic tumors, where the level of coexpression seems to be correlated with the degree of malignancy of the tumor (Pösl et al. 1996).

◄───

Fig. 2. Expression of Bcl-2, Bax, and collagen type X in the growth plate and programmed chondrocyte death in endochondral bone formation. Note the zonal organization (*P*, zone of proliferation; *PH*, prehypertrophic zone; *H*, hypertrophic zone) of maturing chondrocytes within the growth plate (toluidine blue, undecalcified preparation, 5 m). Immunolabeling for collagen X demonstrates characteristic intracellular expression in prehypertrophic cells, whereas the strong pericellular, ring-like matrix labeling shown in *red* is solely found in the hypertrophic zone. Highest levels of Bcl-2 expression are found in late proliferative and prehypertrophic chondrocytes, while Bax expression is increased in hypertrophic chondrocytes (confocal laser microscopy after immunofluorescence labeling with the respective antibodies). Thus, within the growth plate, the ratio of Bcl-2 to Bax progressively decreases in favor of Bax, and fully differentiated chondrocytes consequently die in an apoptotic manner, as confirmed by nick-end labeling of DNA fragments by TdT-mediated dUTP-biotin nick end-labeling (TUNEL)

Fig. 3. Parathyroid hormone-related peptide (PTHrP) increases Bcl-2 and delays chondrocyte differentiation both in vitro and in vivo. Treatment of chondrocytes in vitro with PTHrP shifts the Bcl-2/Bax ratio in favor of Bcl-2 and results in delayed chondrocyte differentiation. The same effect is seen in mice with targeted overexpression of PTHrP to chondrocytes under the control of the col.II promoter. Chondrocytes in these transgenic mice exhibit increased Bcl-2 expression and delayed chondrocyte maturation

Fig. 10. Progressive osteopetrosis in aging *src*-deficient mice. *Text see p. 137*

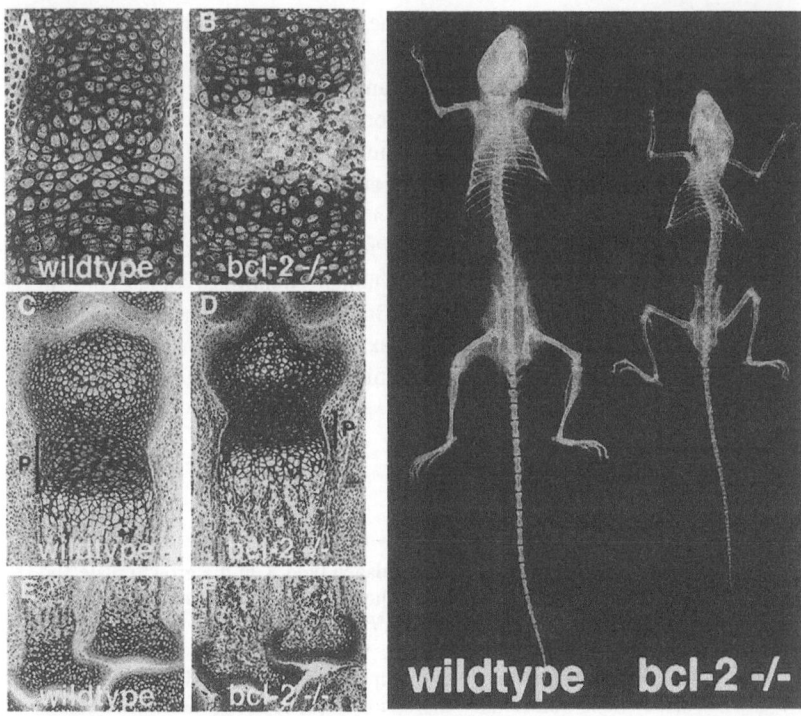

Fig. 4. Bcl-2-deficient mice exhibit accelerated chondrocyte differentiation. Due to the premature maturation of chondrocytes during endochondral bone formation *bcl-2* knock-out mice are consequently markedly smaller than control litter mates. Undecalcified preparation, toluidine blue; contact X-rays

6.3.4 Bmi-1: Skeletal Patterning and Tumorigenesis

Interestingly, the availability of transgenic models has yielded often unexpected insights into the complexity of gene functions.. A recent example from our lab for the convergence of bone tumorigenesis and skeletal patterning is the *bmi-1* proto-oncogene. The human *bmi-1* gene encodes a nuclear protein of 326 amino acids which is homologous to certain members of the Polycomb family of proteins that regulate homeotic gene expression through alteration of the chromatin structure in *Drosophila*. By initially using a differential display approach we identi-

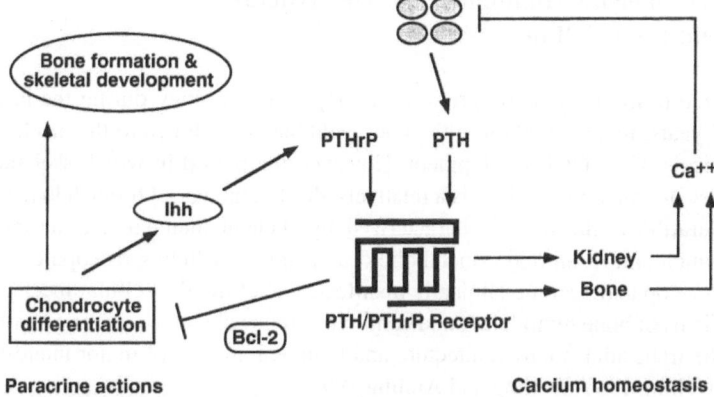

Fig. 5. The control of chondrocyte differentiation and skeletal development

fied *bmi-1* as one of the genes that are overexpressed in high-grade versus low-grade osteosarcoma. Bmi-1 overexpression in osteosarcoma was then further confirmed by western blotting lysates of a variety of primary bone tumors and bone tumor cell lines. Indeed Bmi-1 was found to be specifically overexpressed in osteosarcomas in addition showing a specific speckled subnuclear localization pattern (Hentz et al. 1997). Analyzing animal models, it became clear that Bmi-1 is also involved is skeletal patterning during embryogenesis and manipulation of Bmi-1 results in skeletal phenotypes. Transgenic mice overexpressing Bmi-1 exhibit a dose-dependent anterior transformation of vertebral identity along the complete anteroposterior axis, while at the other end of the spectrum mice with targeted deletion of the *bmi-1* gene show a posterior transformation. This regulation is mediated by repression of specific *hox* genes caused by interaction of Bmi-1 with other members of the mammalian Polycomb complex during development (van der Lugt et al. 1994; Alkema et al. 1997; Hentz et al. 1997). The insights gained from such studies should hasten our understanding of both, skeletal development and tumorigenesis and finally open the way to new forms of therapy.

6.4 Skeletal Maintenance, Bone Structure, and Remodeling

Due to tremendous progress in developmental biology during the last 5 years, much attention in the bone field has been drawn to the mechanisms of skeletal development. However, the period in which skeletal development takes place is a relatively short in humans. Of much longer duration is the period characterized by skeletal maintenance, during which almost all major metabolic osteopathies, including osteoporosis, develop and become clinically manifest. Therefore, the cellular mechanisms of bone remodeling and continuous renewal and reconstruction of the trabecular microarchitecture and bone volume are of major interest (Delling 1987; Delling and Amling 1995).

There are four major targets which may be useful in studying the mechanisms of remodeling responsible for skeletal maintenance: (1) the osteoclast, (2) hormone receptors, (3) bone matrix proteins, and (4) the osteoblast.

6.4.1 Osteoclasts: Osteopetrosis in *op/op* Mice, c-*fos* Knock-Out, and c-*src* Knock-Out Mice

Almost all of the major bone diseases are associated with an increase in osteoclastic bone resorption, which is the primary cause of bone loss. Since the osteoclast is the only cell which is capable of resorbing bone, it is of obvious importance for studying the mechanisms of bone resorption (Baron et al. 1988; Amling and Delling 1996) (Fig. 6).

As another consequence, successful therapies for the most common bone diseases are dependent on our understanding of the molecular mechanisms which regulate osteoclast differentiation and function. Three major animal models have influenced our current knowledge of osteoclast biology. The *op/op* mouse, the c-*fos*–/– mouse, and the *src* –/– mouse, all three presenting with osteopetrosis of varying severity. Both *op/op* mice, which lack circulating colony-stimulating factor-1 (CSF-1) and exhibit reduced numbers of osteoclasts and macrophages, and c-*fos*–/– mice, which lack osteoclasts but not macrophages, demonstrate the essential role of CSF-1 and c-Fos for osteoclastogenesis/osteoclast differentiation. By contrast, *src*–/– mice have even increased numbers of

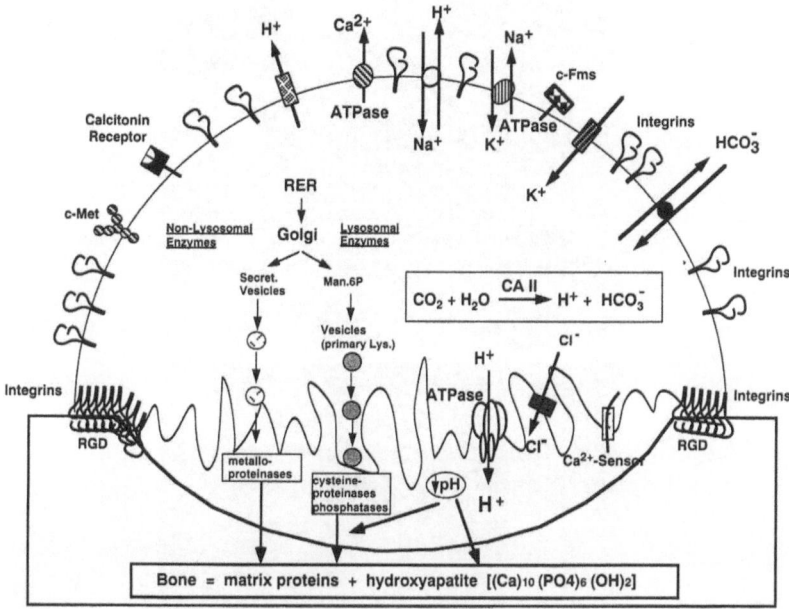

Fig. 6. Mechanisms of osteoclastic bone resorption

osteoclasts, thus osteoclast differentiation does take place in the absence of c-Src; however, *src–/–* osteoclasts are functionally almost inactive and not able to form a ruffled border (Amling et al. 1998) (Figs. 7, 8).

It was Wiktor-Jedrzejczak et al. (1990) who first reported the possibility that a defect in *op/op* mice is due to the failure of hematopoietic stromal cells to release CSF-1 (Fig. 9). Evidence for this was confirmed unequivocally by two independent studies: Yoshida et al. (1990) demonstrated an extra thymidine insertion at base pair 262 in the coding region of the CSF-1 gene in *op/op* mice resulting in a stop codon, TGA, 21 base pairs downstream. At the same time Felix et al. (1990) reported that osteoblastic cells from *op/op* mice could not produce CSF-1 activity. Subsequently, it was reported by Sudas group (Kodama et al. 1991) that administration of recombinant CSF-1 restored the impaired bone resorption of *op/op* mice in vivo.

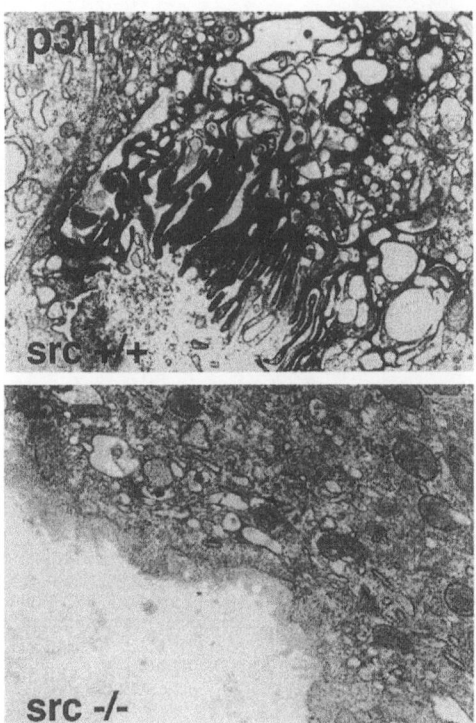

Fig. 7. In the absence of c-Src osteoclasts do not form a ruffled border. Labeling of the p31 subunit of the vacuolar proton pump, which characteristically accumulates along the ruffled border in normal osteoclasts, but not in Src-deficient osteoclasts, is shown

The role of c-Fos as an essential transcription factor for the differentiation of early osteoclast precursors was impressively demonstrated by the phenotype of the c-*fos*–/– mouse (Wang et al. 1992; Grigoriadis et al. 1993, 1994). Moreover the increased number of bone marrow macrophages in c-*fos*–/– mice indicates that c-Fos also affects a related cell type and is perhaps involved in the lineage determination of putative macrophage-osteoclast progenitors (Grigoriadis et al. 1994).

c-Src was first identified as the cellular counterpart of the transforming protein of Rous sarcoma retrovirus, v-Src (Jove and Hanafusa 1987;

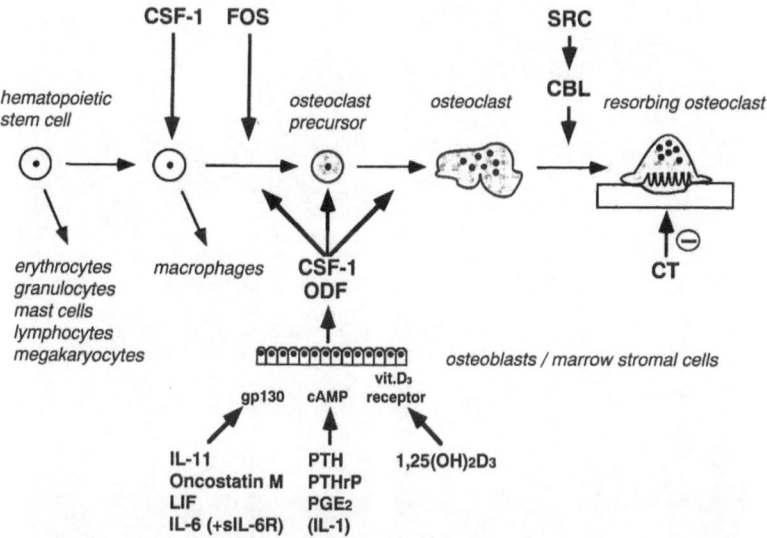

Fig. 8. Osteoclast differentiation and function. *Top*, autocrine regulation; *bottom*, paracrine, osteoblast-mediated regulation due to cytokines and hormones; *right*, direct action of calcitonin on mature osteoclasts

Golden and Brugge 1988). In an effort to elucidate the physiological role of c-Src, Soriano et al. (1991) generated transgenic mice lacking the c-*src* gene. Surprisingly, c-*src*-deficient mice were able to survive and did not show any gross abnormalities in the cell types that were known to express high levels of c-Src, such as platelets and neurons. Unexpectedly, the only phenotype observed in the c-Src-deficient mice was that of osteopetrosis. The skeletal abnormalities in the Src –/– mice include a failure of teeth eruption, increased osteoclast number, the absence of a ruffled border, and in aging mice progressive osteopetrosis and the development of odontomas (Amling et al. 1998) (Fig. 10).

Osteoclasts express high levels of the c-Src protein (Horne et al. 1992; Tanaka et al. 1992) and the defect responsible for the osteopetrotic phenotype of the c-*src*-deficient (*src*–/–) mouse is cell autonomous and occurs in mature osteoclasts (Boyce et al. 1992; Lowe et al. 1993). The specific signaling pathways that require c-Src expression for normal osteoclast activity have, however, not been fully elucidated. We have

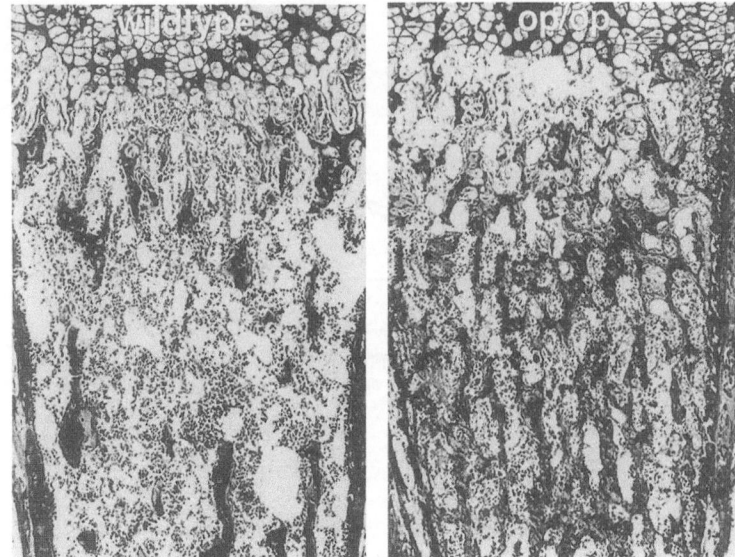

Fig. 9. Osteopetrosis in the *op/op* mouse

Fig. 10. Color plate, see p. 130

shown that the proto-oncogene product c-*Cbl* is tyrosine-phosphory-
lated in a Src-dependent manner in osteoclasts, where the two proteins
colocalize on some vesicular structures (Tanaka et al. 1995, 1996; Blake
et al. 1991). In vitro bone resorption by osteoclast-like cells (OCLs) is
inhibited by both c-*src* and c-*cbl* antisense oligonucleotides (Tanaka et
al. 1996) (Fig. 11). Furthermore, tyrosine phosphorylation of c-*Cbl* and
the localization of c-Cbl-containing structures to the peripheral cytoskeleton are impaired in resorption-deficient *src* –/– OCLs as well as
in wild-type OCLs that have been treated with c-*src* antisense oligonucleotides (Tanaka et al. 1996) (Fig. 12).

Thus, both c-Cbl and c-Src expression are necessary for bone resorption, while c-Src expression is also necessary for c-Cbl phosphorylation.
We therefore conclude that, in osteoclasts, c-Cbl is downstream of c-Src
in a signaling pathway that is required for bone resorption. Although
c-Cbl may not be the only substrate that is not phosphorylated in the

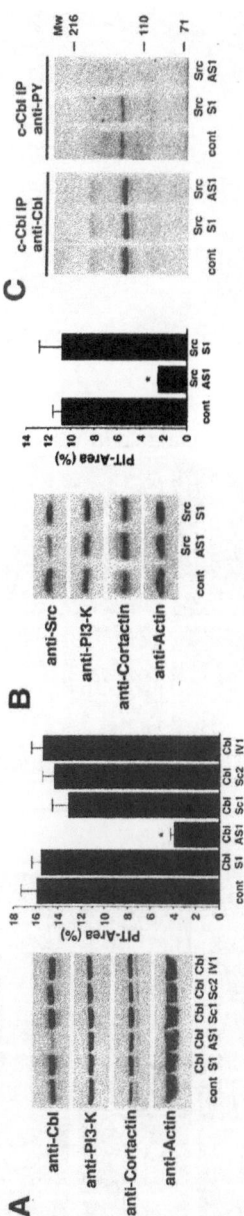

Fig. 11A–C. Src and Cbl are both required for bone resorption and Src is required for Cbl phosphorylation. **A** *cbl* DNA antisense oligonucleotides specifically depleted osteoclasts of c-Cbl and strikingly inhibit bone resorption in the pit assay. **B** *src* DNA antisense oligonucleotides depleted osteoclasts of c-Src and also inhibit bone resorption. **C** In the absence of c-Src tyrosine phosphorylation of Cbl is impaired

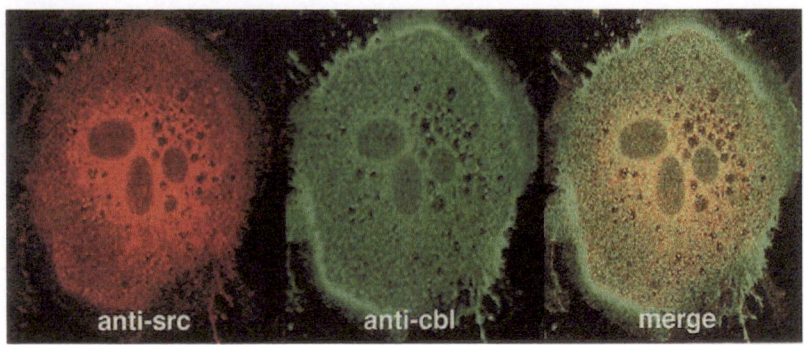

Fig. 12. Legend see p. 141

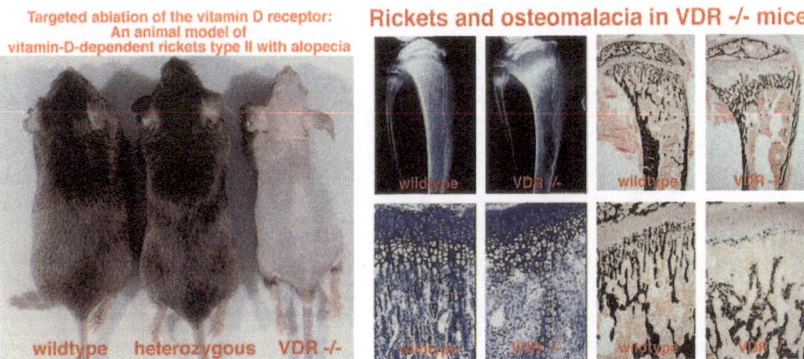

Fig. 13. Legend see p. 141

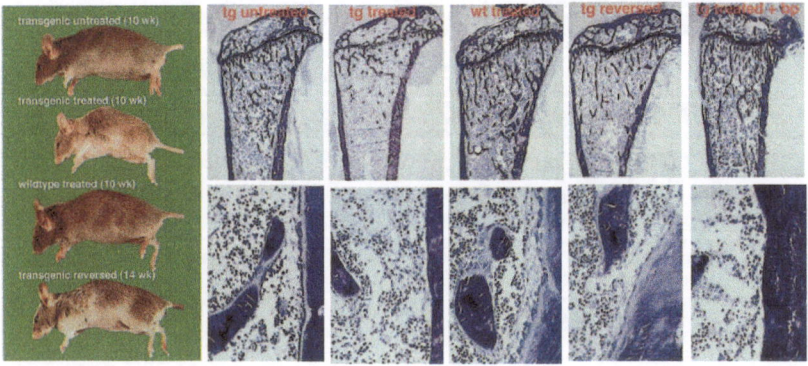

Fig. 15. Legend see p. 141

absence of c-Src, the disruption of this pathway may be involved in the osteopetrotic phenotype of the *src* –/– transgenic mouse, leaving other cells not detectably affected (Tanaka et al. 1996).

6.4.2 Hormone Receptors: Rickets Type II with Alopecia in Vitamin D Receptor Knock-Out Mice

1, 25 dihydroxyvitamin D is the major steroid hormone of mineral ion homeostasis. Its actions are thought to be mediated by a nuclear receptor, the vitamin D receptor (VDR), which heterodimerizes with the retinoid X receptor and interacts with specific DNA sequences on target genes. The VDR is evolutionarily well conserved and is expressed early in development in amphibians, mammals, and birds. As well as being expressed in the intestine, the skeleton, and the parathyroid glands, the VDR is found in several tissues not thought to play a role in mineral ion homeostasis. Its precise functions in these tissues, as well as its developmental role remain unclear. Although the VDR is widely expressed early during embryonic development, no major developmental abnormalities are observed in the VDR knock-out mice or in humans with vitamin D dependent rickets type II (VDDR II). VDR knock-out animals are normal at birth, however, they develop hypocalcemia, hyperparathyroidism, and alopecia within the first month of life (Li et al. 1997) (Fig. 13).

Although VDR –/– animals are normocalcemic until day 21, they become progressively hypocalcemic after that point. Concomitant with hypocalcemia, a progressive increase in serum immunoreactive PTH levels was observed from day 21 in VDR –/– mice, and the animals became hypophosphatemic by day 21. The time of the onset of the

◄───

Fig. 12. Subcellular localization of c-Src and c-Cbl in the osteoclast. Confocal microscopy after immunofluorescence double labeling for Src (*red*) and Cbl (*green*)

Fig. 13. Rickets, osteomalacia, and alopecia in vitamin D receptor (VDR) –/– mice

Fig. 15. Reversible ablation of osteoblasts in OG2 HSV-TK mice. *Text see p. 144*

hypocalcemia in the VDR–/– mice is not unexpected in view of the observation that intestinal calcium absorption in rats occurs by a non-saturable 1, 25 vitamin D-independent mechanism the first 18 days of life (Dostal and Toverud 1984). Interestingly, the growth plate abnormalities in VDR -/- mice precede the development of disordered mineral ion homeostasis, which is observed as early as 15 days of age. These data suggest that, although the receptor-dependent actions of 1, 25 dihydroxyvitamin D are not necessary for normal embryogenesis, they may play a role in the maturation of chondrocytes during longitudinal growth, even in the setting of normal mineral ion homeostasis. By 5 weeks of age, however, both a lack of osteoid mineralization (15-fold increased osteoid volume) and profound abnormalities in the growth plate were observed.

Interestingly the phenotype also demonstrated that, in the setting of a whole animal, the VDR is not indispensable for osteoclast differentiation and function, as functionally active osteoclasts were detectable on the bone surface of the knock-out animals. The critical role of calcium is further documented by the fact that the phenotype could drastically be shifted toward normal by a high calcium diet. An independent study (Yoshizawa et al. 1997) was entirely consistent with these findings, leaving the data on the shorter life span of the Japanese mice unaddressed.

6.4.3 Bone Matrix Proteins: Different Effects on Bone Volume of Osteonectin and Osteocalcin

Bone consists of matrix proteins and the cells that make, mineralize, resorb, renew, and maintain them. The matrix of bone, which is physiologically mineralized with hydroxyapatite, is composed of a multitude of proteins that determine its unique characteristics and functions. While 90% of total bone proteins consists of type I collagen, noncollagenous proteins (NCP) account for the other 10% of bone protein. However, the physiological roles of most individual bone proteins remain undefined. Recently knock-out experiments shed some light on at least two of the most abundant NCPs: osteocalcin (bone gla-protein) and osteonectin/ SPARC (secreted protein, acid and rich in cysteine).

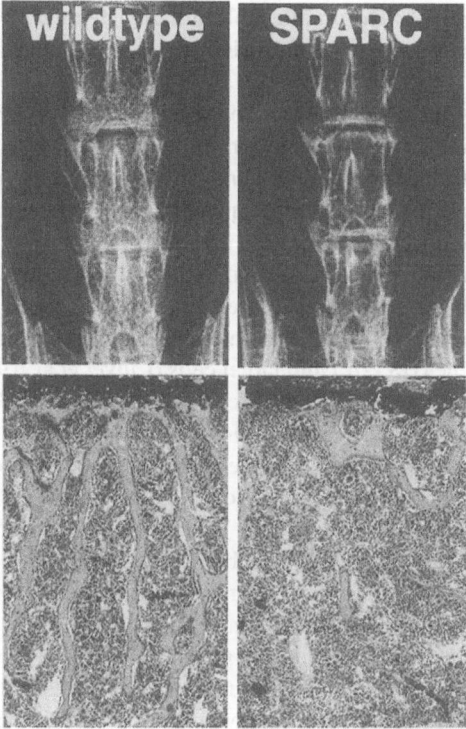

Fig. 14. Osteopenia in the osteonectin (SPARC)-deficient mouse

Last year Karsenty and coworkers (Ducy et al. 1996) reported that osteocalcin-deficient mice develop a phenotype marked by higher bone mass. This suggested that osteocalcin is a negative regulator of bone formation without impairing bone resorption or mineralization. The molecular mechanism by which osteocalcin controls bone matrix deposition remain however unknown.

Early results in SPARC-deficient mice suggest that in the absence of osteonectin mice develop osteopenia (Fig. 14). These findings together indicate that a further understanding of the biological role of these abundant bone matrix proteins is needed to clarify the complex regulatory pathways of bone maintenance.

6.4.4 Osteoblasts: Reversible Ablation of Osteoblasts:
An Animal Model of Osteoporosis

One assumption of the theory of bone metabolic units (BMUs) is that bone formation and bone resorption are mechanistically coupled during skeletal maintenance and remodeling. However the existence of a functional link between bone formation and bone resorption has never been demonstrated conclusively in vivo. To define the role of bone formation in the regulation of bone resorption in vivo, we generated an inducible osteoblast ablation model. We used an emerging strategy for cancer gene therapy which involves the transfer of the herpes simplex thymidine kinase gene (HSV-*tk*) into target cells (Culver et al. 1992; Hamel et al. 1996). Transgenic mice were generated in which a 1.3 kb fragment of the osteocalcin gene 2 (OG2) drove expression of the HSV-*tk*. The OG2 promoter is sufficient to achieve osteoblast-specific expression of HSV-TK in vivo. Since dividing cells expressing HSV-TK die upon treatment with gancyclovir (GCV), HSV-TK expression in dividing osteoblasts allows inducible osteoblast ablation in vivo. In transgenic mice, osteoblast ablation leads to an arrest of skeletal growth and to the development of osteopenia. Serum levels of osteocalcin are dramatically decreased, while calcium and phosphate levels remain unchanged. Histologically the bones were denuded of osteoblasts and bone formation rate was zero. Upon withdrawal of GCV there was a complete reversal of the phenotype. Most interestingly, the number of osteoclasts remained unchanged and the bone volume was decreased after osteoblast ablation. Indeed in the absence of bone formation bone resorption occurred both in vivo and in vitro. These results indicate clearly that bone resorption is not controlled by, and not coupled to, bone formation. Furthermore, this animal model is amenable to modulation in respect to the severity of the phenotype. In addition bone resorption can be maintained in the absence of bone formation for even longer periods of time. Consequently OG2 HSV-TK mice can be used to mimic osteoporosis of varying degrees of severity by continuing bone resorption in the face of little or no bone formation (Corral et al. 1997) (Fig. 15).

Fig. 15. Color plate, see p. 140

This animal model provides a new tool that can be used to address several questions regarding osteoporosis that could not be addressed previously, including for instance the role of peak bone mass, the efficacy of antiresorptive drugs, and the feasibility of novel approaches to treatment of osteoporosis, e.g., gene therapy.

6.5 The Genetic Perspective

Although this chapter focuses on the bone-specific issues of gene targeting, the general perspective should be mentioned briefly. Protocols using transgenesis (Gordon et al. 1980) and homologous recombination in embryonic stem cells (ES) (Capecchi 1989a, b) permit inactivation, overexpression and modification of genes almost at will. These technologies are invaluable in assessing the role of genes in complex processes such as development, tumorigenesis, and cell signaling and function. However, there are several limitations, beside the above mentioned fact that mice with the same genetic mutation do not always mimic the human symptoms. For example nullizygosity appears to be lethal in many instances or causes complex pleiotropic effects and therefore does not permit the development of an in vivo model system in which gene inactivation is restricted to a defined subset of cells (Copp 1995).

To overcome these limitations, strategies for conditional, cell type-specific gene targeting (Gu et al. 1994), inducible gene disruption (Kühn et al. 1995), and cell type specific ablation have recently been developed. Some of these systems take advantage of site-specific recombinases such as the Cre/loxP recombination system of bacteriophage P1 (Fig. 16). Only two components are required: the 38 kDa Cre (causes recombination) recombinase from bacteriophage P1, which belongs to the integrase family of recombinases. Cre catalyzes site-specific recombination between specific DNA target sites of 34 bp each, termed loxP (locus of crossing over) (reviewed by Plück 1996). The utility of this system has been shown by both the generation of conditional transgenic mice and the production of conditional gene knock-outs (St-Onge et al. 1996; Mullins et al. 1997). In the latter case, a target construct flanked by two loxP sites (flox) was used to modify the cognate gene by homologous recombination in ES cells. The expression level of the floxed allele is expected to be the same as that of the wild-type and

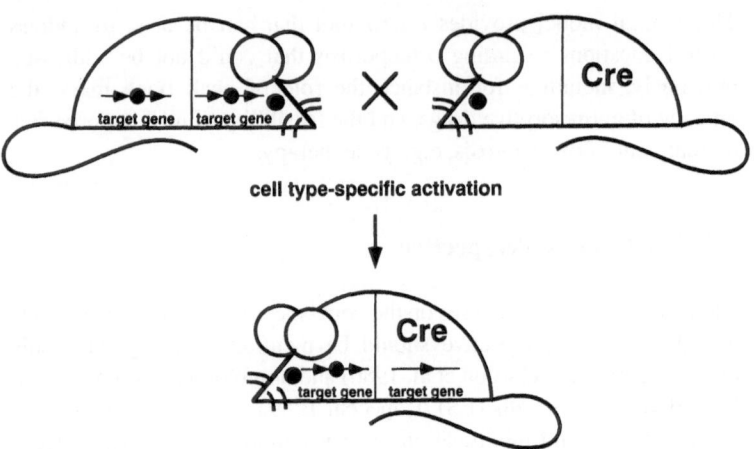

Fig. 16. Cre/loxP mice. Breeding scheme for the approach of cell type-specific gene targeting by homologous and site-specific recombination. *Upper left,* a mouse in which a gene in the germ line (*black dot*) is flanked by two loxP sequences (*triangles*); *upper right,* a Cre transgenic mouse; *bottom,* the target gene is deleted in the tissue in which the Cre enzyme is expressed

should therefore not lead to phenotypic changes. Crossing of the floxed mice with transgenic mice carrying the Cre recombinase gene under the control of a cell type-specific promoter or an inducible promoter then leads to excision of the intervening sequences. This strategy has been shown to work in a number of settings (Kuhn et al. 1995; Kilby et al. 1993; Lakso et al. 1992). This new technology is a milestone in the field of mouse reverse genetics and will have significant impact on skeletal research. However, full exploitation of this system requires further improvements including, e.g., the use of adeno-Cre viruses (Akagi et al. 1997). In this respect the control of Cre expression appears critical. Problems to be solved are tissue specificity of expression, background activity, level of induction, control over fraction of cells in which expression can be induced and the timing of expression.

Recent studies of human and mice with skeletal defects (dysplasias, metabolic disorders, and tumors) have pointed to many genes important in skeletal development and skeletal maintenance. Our understanding of skeletal morphology is starting to extend by insights into the molecular mechanisms controlling bone cell differentiation and function. The questions which can be addressed by further development of strategies like the Cre/loxP system are stimulating novel forms of genetic analyses in bone. Our future progress towards a better understanding of bone physiology will depend on the successful convergence of these novel approaches with established and accurate knowledge of bone pathology. Indeed histology, endocrinology, histomorphometry, cell biology, and genetics together will lead to the development of new therapeutic strategies for major bone diseases.

6.6 Summary

Today our morphological understanding of the skeleton is increasing due to insights into the molecular mechanisms controlling bone cell differentiation and function. Genetic analyses and recent studies of human and mice with skeletal defects have pointed to many genes important in skeletal development and skeletal maintenance. Among mammals, mice are the most promising animals for these kinds of experiments. Since there is extensive genetic information available, many mouse mutations are known, and cells from early mouse developmental stages are accessible. Scientists have developed transgenic mice – in which a gene is introduced or ablated in the germ line. Thus far, we have analyzed some 60 different transgenic and knock-out models with various skeletal phenotypes, covering the major aspects of both skeletal development and skeletal maintenance. Based on these results we have presented an early perspective on transgenic and gene knock-out animals in skeletal research, including insights into signaling pathways controlling endochondral bone formation, the regulation of osteoblast function, and osteoclastic bone resorption, as well as bone tumorigenesis. Furthermore, these data demonstrate that the successful convergence of novel genetic approaches with the established and fundamental knowledge of bone pathology has made a beginning. A wealth of detail about the skeletal system is available. Still, the successes do not amount to a complete or even very profound understanding. On the contrary, our current ignorance is vaster than our current knowledge and there remains to be discovered mechanisms and concepts that no one has yet even imagined. In some instances, we have learned enough to at least identify important areas of ignorance. However, the challenges are great, and the use of transgenic mice to dissect and analyze regulatory mechanisms in bone cell physiology and the pathogenesis of human bone diseases remains an extremely powerful experimental tool. Indeed, we can be certain of one thing: from histology, endocrinology, histomorphometry, cell biology, and genetics together will emerge major new concepts in skeletal biology. These will lead to the development of new therapeutic strategies for treatment of some of the major bone diseases.

Acknowledgements. The authors would like to thank the members of the Delling laboratory, especially Marc W. Hentz, and Matthias Priemel, for critical discussions, intellectual curiosity, and technical help. M.A. thanks Dr. Günter Delling for tremendous personal support and encouragement during the course of these studies. We are grateful to all colleagues and collaborators for sharing unpublished data. We are especially grateful to our collaborators Drs. Lynn Neff, Roland Baron (Yale University), Sakae Tanaka (Tokyo University), Daisuke Inoue (Tokushima University), Patricia Ducy, Gerard Karsenty (Baylor College, Houston), Marie Demay, Beate Lanske, Hank Kronenberg (Harvard University), Anne Delany (Hartford), Douglas Coffin (Great Falls), Maarten van Lohuizen (The Netherlands Cancer Institute), and Phillippe Clement-Lacroix (INSERM Paris).

References

Akagi K, Sandig V, Vooijs M, Van der Valk M, Giovannini M, Strauss M, Berns A (1997) Cre-mediated somatic site-specific recombination in mice. Nucleic Acids Res 25:1766–1773

Alkema MJ, Bronk M, Verhoeven E, Otte A, van't Veer LJ, Berns A, van Lohuizen M (1997) Identification of Bmi-1-interacting proteins as constituents of a multimeric mammalian polycomb complex. Genes Dev 11:226–240

Allsopp TE, Wyatt S, Paterson HF, Davies AM (1993) The proto-oncogene bcl-2 can selectively rescue neurotrophic factor-dependent neurons from apoptosis. Cell 73:295–307

Amizuka N, Warshawsky H, Henderson JE, Goltzman D, Karaplis AC (1994) Parathyroid hormone-related peptide-depleted mice show abnormal epiphyseal cartilage development and altered endochondral bone formation. J Cell Biol 126:1611–123

Amizuka N, AC Karaplis, Henderson JE, Warshawsky H, Lipman ML, Matsuki Y, Ejiri S, Tanaka M, Izumi N, Ozawa H, Goltzman D (1996a) Haploinsufficiency of parathyroid hormone-related peptide (PTHrP) results in abnormal postnatal bone development. Dev Biol 175:166–176

Amizuka N, Henderson JE , Hoshi K, Warshawsky H, Ozawa H, Goltzman D, Karaplis AC (1996b) Programmed cell death of chondrocytes and aberrant chondrogenesis in mice homozygous for parathyroid hormone-related peptide gene deletion. Endocrinology 137:5055–5067

Amling M, Delling G (1996) Zellbiologie des Osteoklasten und molekulare Mechanismen der Knochenresorption Pathologe 17: 358–367

Amling M, Grote H, Pösl M, Hahn M, Delling G (1994) Polyostotic heterogeneity of the spine in osteoporosis. Bone Mineral 27: 193–208

Amling M, Herden S, Pösl M, Hahn M, Ritzel H, Delling G (1996) Heterogeneity of the skeleton. J Bone Miner Res 11: 36–45

Amling M, Neff L, Tanaka S, Inoue D, Kuida K, Weir E C, Philbrick WM, Broadus AE, Baron R (1997) Bcl-2 lies downstream of parathyroid hormone-related peptide in a signaling pathway that regulates chondrocyte maturation during skeletal development. J Cell Biol 136:205–213

Amling M, Neff L, Priemel M, Delling G, Baron R (1998) Progressive osteopetrosis and development of odontomas in aging Src-deficient mice. Bone (in press)

Baker J, Liu JP, Robertson EJ, Efstratiadis A (1993) Role of insulin-like growth factors in embryonic and postnatal growth. Cell 75:73–82

Baron R, Neff L, Brown W, Courtoy PJ, Louvard D, Farquhar MG (1988) Polarized secretion of lysosomal enzymes: co-distribution of cation-independent mannose-6-phosphate receptors and lysosomal enzymes along the osteoclast exocytic pathway. J Cell Biol 106:1863–1872

Blake TJ, Shapiro M, Morse HC, Langdon WY (1991) The sequence of the human c-cbl proto-oncogene show v-cbl was generated by a large truncation encompassing a proline-rich domain and a leucine zipper-like motif. Oncogene 6:653–657

Boise LH, González-García M, Postema CE, Ding L, Lindsten T, Turca LA, Mao X, Nunez G, Thompson CB (1993) bcl-x, a bcl-2-related gene that functions as a dominant regulator of apoptotic cell death. Cell 74:597–608

Boyce BF, Yoneda T, Lowe C, Soriano P, Mundy GR(1992) Requirement of pp60c-src expression for osteoclasts to form ruffled borders and resorb bone in mice. J Clin Invest 90:1622–1627

Broadus A, Stewart E AF (1994) Parathyroid hormone-related protein. In Bilezikian JP, Levine MA, Marcus R (eds) The parathyroids. Raven, New York, pp 259–294

Capecchi MR (1989a) The new mouse genetics: altering the genome by gene targeting. Trends Genet 5:70–76

Capecchi MR (1989b) Altering the genome by homologous recombination. Science 244:1288–1292

Coffin JD, Florakiewicz RZ, Neumann J, Mort-Hopkins T, Dorn II GW, Lightfoot P, German R, Howles PN, Kier A, O'Toole BA, Sasse J, Gonzales AM, Gonzales Baird A, Doetschman T (1995) Abnormal bone growth and selective translational regulation in the basic fibroblast growth factor (FGF-2) transgenic mice. Mol Biol Cell 6:1861–1873

Copp AJ (1995) Death before birth: clues from gene knock-out and mutations. Trends Genet 11:87–93

Corral D, Amling M, Priemel M, Neff L, Chung U, Kronenberg HM, Baron R, Karsenty G (1997) Reversible ablation of osteoblasts: An animal model of osteoporosis. J Bone Min Res 12: S120

Culver KW, Ram Z, Wallbridge S, Ishii H, Oldfield EH, Blaese RM (1992) In vivo gene transfer with retroviral vector-producer cells for treatment of experimental brain tumors. Science 256:1550–1552

Delling G (1987) Bone morphology in primary hyperparathyroidism A qualitative and quantitative study of 391 cases Appl Pathol 5: 157–159

Delling G, Amling M (1995) Biomechanical stability of the skeleton – it is not only bone mass, but also bone structure that counts. Nephrol Dial Transplant 10: 601–606

Deng C, Wynshaw-Boris A, Zhou F, Kuo A, Leder P (1996) Fibroblast growth factor receptor 3 is a negative regulator of bone growth. Cell 84:911–921

Dostal LA, Toverud SU (1984) Effect of vitamin D3 on duodenal calcium absorption in vivo during early development. Am J Physiol 246:G528–G534

Ducy P, Desbois C, Boyce B, Pinero G, Story B, Dustan C, Smith E, Bonadio J, Goldstein S, Grundberg C, Bradley A, Kasenty G (1996) Increased bone formation in osteocalcin-deficient mice. Nature 382:448–452

Ducy P, Zhang R, Geoffroy V, Ridall AL, Karsenty G (1997) Osf2/Cbfa1: a transcriptional activator of osteoblast differentiation. Cell 89:747–745

Erlebacher A, Filvaroff EH, Gitelman SE, Derynck R (1995) Towards a molecular understanding of skeletal development. Cell 80:371–378

Felix R, Cecchini MG, Fleisch H (1990) Macrophage colony stimulating factor restores in vivo bone resorption in the op/op osteopetrotic mouse. Endocrinology 127:2592–2594

Golden AJ, Brugge S (1988) The src oncogene. In: Reddy EP, Skalka AM, Curran T (eds) The oncogene handbook. Elsiver, Amsterdam, pp 149–173

Goltzman D, Hendy GN, Banville D(1989) Parathyroid hormone-like peptide: molecular characterization and biological properties. Trends Endocrinol Metabol 1:39–44

Gordon JW, Scangos GA, Plotkin DJ, Barbosa JA, Ruddle FH (1980) Genetic transformation of mouse embryos by microinjection of purified DNA. Proc Natl Acad Sci USA 77:7380–7384

Grigoriadis AE, Schellander K, Wang ZQ, Wagner EF (1993) Osteoblasts are target cells for transformation in c-fos transgenic mice. J Cell Biol 122:685–701

Grigoriadis AE, Wang ZQ, Cecchini MG, Hofstetter W, Felix R, Fleisch H, Wagner E F (1994) c-Fos: A key regulator of osteoclast-macrophage lineage determination and bone formation. Science 266:443–448

Gu H, Marth JD, Orban PC, Mossmann H, Rajewsky K (1994) Deletion of dna polymerase b gene segment in t cells using cell type-specific gene targeting. Science 265:103–105

Hamel W, Magnelli L, Chiarugi VP, Israel MA (1996) Herpes simplex virus thymidine kinase/ganciclovir-mediated apoptotic death of bystander cells. Cancer Res 56:2697–2702

Hardy MH (1992) The secret life of the hair follicle. Trends Genet 8:55–61

Hentz MW, Amling MJ, David P, Neff L, van Lohuizen M, Baron R, Delling G (1997) Overexpression and altered subnuclear localization of the transcriptional repressor Bmi-1 in osteosarcoma. J Bone Miner Res 12:S140

Hockenbery DM, Oltvai ZN, Yin XM, Milliman CL, S J Korsmeyer (1993) Bcl-2 functions in an antioxidative pathway to prevent apoptosis. Cell 75:241–251

Horne WC, Neff L, Lomri A, Levy JB, Baron R (1992) Osteoclasts express high levels of pp60c-src in association with intracellular membranes. J Cell Biol 119:1003–1013

Jansen M (1934) Über atypische Chondrodystrophie (Achondroplasie) und über eine noch nicht beschriebene angeborene Wachstumsstörung des Knochensystems: Metaphysäre Dysostosis. Z Orthop Chir 61:253–286

Jensen BL (1990) Somatic development in cleidocranial dysplasia. Am J Med Gen 35:69–74

Jove R, Hanafusa H (1987) Cell transformation by the viral src oncogene. Ann Rev Cell Biol 3:31–56

Jüppner H (1996) Jansen's metaphyseal chondrodysplasia. A disorder due to a PTH/PTHrP receptor gene mutation. Trends Endocrinol Metabol 7:157–162

Jüppner H, Abou-Samra AB, Freeman M, Kong XF, Schipani E, Richards J, Kolakowski LF, Hock J, Potts JT, Kronenberg HM, Segre GV (1991)A G protein-linked receptor for parathyroid hormone and parathyroid hormone-related peptide. Science 254:1024–1026

Jüppner H, Abou-Samra AB, Uneno S, Gu WX, Potts JT, Segre GV (1988) The parathyroid hormone-like peptide associated with humoral hypercalcemia of malignancy and parathyroid hormone bind to the same receptor on the plasma membrane of ROS 17/28 cells. J Biol Chem 263:8557–8560

Karaplis AC, Luz A, Glowacki J, Bronson RT, Tybulewicz VLJ, Kronenberg HM, Mulligan RC (1994) Lethal skeletal dysplasia from targeted disruption of the parathyroid hormone-related peptide gene. Genes Dev 8:277–289

Kilby NJ, Snaith MR, Murray JA (1993) Site specific recombinase: tools for genome engineering. Trends Genet 9:413–421

Kodama H, Yamasaki A, Nose M, Niida S, Ohgame Y, Abe M, Kumegawa M, Suda T (1991)Congenital osteoclast deficiency in osteopetrotic (op/op) mice is cured by injections of macrophage colony-stimulating factor. J Exp Med 173:269–274

Komori T, Yagi H, Nomura S, Yamaguchi A, Sasaki K, Deguchi K, Shimizu Y, Bronson RT, Gao YH, Inada M, Sato M, Okamoto R, Kitamura Y, Yoshiki S, Kishimoto T (1997) Targeted disruption of CBFA 1 results in a complete lack of bone formation owing to maturational arrest of osteoblasts. Cell 89:755–764

Korsmeyer SJ (1992) Bcl-2 initiates a new category of oncogenes: regulators of cell death. Blood 80:879–886

Kronenberg HM, Bringhurst FR, Nussbaum S, Jüppner H, Abou-Samra A-B, Segre G, Potts JT (1993) Parathyroid hormone: Biosynthesis, secretion, chemistry, and action. In: Mundy GR, Martin TJ (eds) Handbook of experimental pharmacology: physiology and pharmacology of bone. Springer-Verlag, Berlin, Heidelberg, New York, pp 507–549

Kühn R, Schwenk F, Aguet M, Rajewsky K (1995) Inducible gene targeting in mice. Science 269:1427–1429

Lakso M, Sauer B, Mosinger BJ, Lee EJ, Manning RW, Yu SH, Mulder KL, Westphal H (1992) Targeted oncogene activation by site specific recombinases in transgenic mice. Proc Natl Acad Sci USA 89:6232–6236

Lanske B, Karaplis AC, Lee K, Luz A, Vortkamp A, Pirro A, Karperien M, Defize LHK, Ho C, Mulligan RC, Abou-Samra AB, Jüppner H, Segre GV, Kronenberg HM (1996) PTH/PTHrP receptor in early development and Indian hedgehog-regulated bone growth. Science 273:663–666

Li Y, Lacerda DA, Warman ML, Beier ML, Yoshioka H, Ninomiya Y, Oxford JT, Morris NP, Andrikopoulos K, Ramirez F, Wardell BB, Lifferth GD, Teuscher C, Woodward SR, Tayler BA, Seegmiller RE, Olsen BR (1995) A fibrillar collagen gene, Col11a1, is essential for skeletal morphogenesis. Cell 80:423–430

Li YC, Pirro AE, Amling M, Delling G, Baron R, Bronson R, Demay MB (1997) Targeted ablation of vitamin D receptor: an animal model of vitamin d-dependent rickets type II with alopecia. Proc Natl Acad Sci USA 94:9831–9835

Lowe C, Yoneda T, Boyce BF, Chen H, Mundy GR, Soriano P (1993) Osteopetrosis in src-deficient mice is due to an autonomous defect of osteoclast. Proc Natl Acad Sci USA 90:4485–4489

Mangin M, Webb AC, Dreyer BE, Posillico JT, Ikeda K, Weir E C, Stewart AF, Bander NH, Milstone L, Barton DE, Francke U, Broadus AE (1988) Identification of a cDNA encoding a parathyroid hormone-like peptide from human tumor associated with humoral hypercalcemia of malignancy. Proc Natl Acad Sci USA 85:597–601

Muenke M, Schell U (1995) Fibroblast-growth-factor receptor mutations in human skeletal disorders. Trends Genet 11:308–313

Mullins LJ, Kotelevtseva N, Boyd AC, Mullins JJ (1997) Efficient cre-lox linearisation of bacs: applications to physical mapping and generation of transgenic animals. Nucleis Acids Res 25:2539–2540

Mundlos S, Otto F, Mundlos C, Mulliken JB, Aylsworth AS, Albright S, Lindhout D, Cole WG, Henn W, Knoll JHM, Owen MJ, Mertelsmann R, Zabel BU, Olsen BR (1997) Mutations involving the transcription factor CBFA 1 cause cleidocranial dysplasia. Cell 89:773–779

Nakayama K, Nakayama K-I, Negishi I, Kuida K, Sawa H, Loh DY (1994) Targeted disruption of Bcl-2ab in mice: occurence of gray hair, polycystic kidney disease, and lymphocytopenia. Proc Natl Acad Sci USA 91:3700–3704

Ogawa E, Maruyama M, Kagoshima H, Inuzuka M, Lu J, Satake M, Shigesada K, Ito Y (1993) PEBP2/PEA2 represents a family of transcription factors homologous to the products of Drosophila runt gene and the human AML1 gene. Proc Natl Acad Sci USA 90:6859–6863

Oltvai ZN, Milliman CL, Korsmeyer SJ (1993) Bcl-2 heterodimerizes in vivo with a conserved homolog, Bax, that accelerates programmed cell death. Cell 74:609–6(19

Otto F, Thornell AP, Crompton T, Dencel A, Gilmour KC, Rosewell IR, Stamp GWH, Beddington RSP, Mundlos S, Olsen BR, Selby PB, Owen MJ (1997) Cbfa 1, a candidate gene for cleidocranial dysplasia syndrome, is essential for osteoblast differentiation and bone development. Cell 89:765–771

Plück A (1996) Conditional mutagenesis in mice: the cre/loxP recombination system. Int J Exp Path 77:269–278

Pösl M, Amling M, Neff L, Grahl K, Baron R, Delling G (1996) Coexpression of PTHrP and Bcl-2 correlates with the degree of malignancy of chondrogenic tumors. J Bone Miner Res 11:S113

Reimhold AM, Grusby MJ, Kosaras B, Fries JWU, Mori R, Maniwa S, Clauss IM, Collins T, Sidam RL, Glimcher M, Glimcher LH (1996) Chondrodysplasia and neurological abnormalities in the ATF-2-deficient mice. Nature 379:262–265

Rousseau F, Bonaventure J, Legeal-Mallet L, Pelet A, Rozet J-M, Maroteaux P, Le Merrer M, Munnich A (1994) Mutations in the gene encoding fibroblast growth factorreceptor-3 in achondroplasia. Nature 371:252–254

Rousseau F, Saugier P, Le Merrer M, Munnich A, Delezoide A-L, Maroteaux P, Bonaventure J, Narcy F, Sanak M (1995) Stop codon FGFR3 mutations in thanatophoric dwarfism type 1. Nat Genet 10:11–12

Schipani E, Kruse K, Jüppner H (1995) A constitutively active mutant PTH-PTHrP receptor in Jansen-type metaphyseal chondrodysplasia. Science 268:98–100

Shiang R, Thompson LM, Zhu Y-Z, Church DM, Fielder TJ, Bocian M, Winokur ST, Wasmuth JJ (1994) Mutations in the transmembrane domain of FGFR3 cause the most common genetic form of dwarfism, achondroplasia. Cell 78:335–342

Snell GD (1929) Dwarf, a new mendelian recessive character of the house mouse. Proc Natl Acad Sci USA 15:733–734

Soriano P, Montgomery C, Geske R, Bradley A (1991)Targeted disruption of the c-src proto-oncogene leads to osteopetrosis in mice. Cell 64:693–702

St-Onge L, Furth PA, Gruss P (1996) Temporal control of the cre recombinase in transgenic mice by a tetracycline responsive promoter. Nucleic Acids Res 24:3875–3877

Strewler GJ, Stern PH, Jacobs JW, Eveloff J, Klein RF, Leung SC, Rosenblatt M, Nissenson RA (1987) Parathyroid hormone-like protein from human renal carcinoma cells Structural and functional homology with parathyroid hormone. J Clin Invest 80:1803–1807

Suva LJ, Winslow GA, Wettenhall RE H, Hammonds RG, Moseley JM, Diefenbach-Jagger H, Rodda CP, Kemp BE, Rodriguez H, Chen E Y, Hudson PJ, Martin TJ, Wood WI (1987) A parathyroid hormone-related protein implicated in malignant hypercalcemia: cloning and expression. Science 237:893–896

Tanaka S, Amling M, Neff L, Peymann A, Uhlmann E, Levy JB, Baron R (1996) c-Cbl is downstream of c-Src in a signaling pathway necessary for bone resorption. Nature 383:528–531

Tanaka S, Neff L, Baron R, Levy JB (1995) Tyrosine phosphorylation and translocation of the c-Cbl protein after activation of tyrosine kinase signaling pathways. J Biol Chem 270:14347–14351

Tanaka S, Takahashi N, Udagawa N, Sasaki T, Fukui Y, Kurokawa T, Suda T (1992 Osteoclasts express high levels of p60c-src, preferentially on ruffled border membranes. FEBS Lett 313:85–89

van der Lugt NMT, Domen J, Linder K, van Roon M, Robanus-Maandag E, te Riele H, Van der Valk M, Deschamps J, Sofroniew M, van Lohuizen M, Berns A (1994) Posterior transformation, neurological abnormalities, and severe hematopoietic defects in mice with targeted deletion of the bmi-1 proto-oncogene. Genes Dev 8:757–769

van de Stolpe A, Karperien M, Lšwik CWGM, Jüppner H, Segre GV, Abou Samra A-B, de Laat SW, Defize LHK (1993) Parathyroid hormone-related peptide as an endogenous inducer of parietal endoderm differentiation. J Cell Biol 120:235–243

Veis DJ, Sorenson CM, Shutter JR, Korsmeyer SJ (1993) Bcl-2-deficient mice demonstrate fulminant lymphoid apoptosis, polycystic kidneys, and hypopigmented hair. Cell 75:229–240

Vikkula M, Mariman E CM, Lui VCH, Zhidkova NI, Tiller GE, Goldring MB, van Beersum SE C, de Waal Malefijt M, van den Hoogen FHJ, Ropers HH, Mayne R, Cheah KSE, Olsen BR, Warman ML, Brunner HG (1995) Autosomal dominant and recessive osteochondrodysplasias associated with the Col11A2 locus. Cell 80:431–437

Vortkamp A, Lee K, Lanske B, Segre GV, Kronenberg HM, Tabin CJ (1996) Regulation of rate of cartilage differentiation by indian hedgehog and PTH-related protein. Science 273:613–622

Wang Z-Q, Ovitt C, Grigoriadis AE, Möhle-Steinlein U, Rüther U, Wagner E F (1992 Bone and haematopoietic defects in mice lacking c-fos. Nature 360:741–745

Weir E C, Philbrick WM, Amling M, Neff L, Baron R, Broadus AE (1996) Targeted overexpression of parathyroid hormone-related peptide in chondrocytes delays chondrocyte differentiation and endochondral bone formation. Proc Natl Acad Sci USA 93: 10240–10245

Wiktor-Jedrzejczak W, Bartocci A, Ferrante AW, Ahmed-Ansari A, Sell KW, Pollard JW, Stanley E R (1990) Total absence of colony-stimulating factor 1 in the macrophage-deficient osteopetrotic (op/op) mouse. Proc Natl Acad Sci USA 87:4828–4832

Wysolmerski JJ, Broadus AE, Zhou J, Fuchs E, Milstone LM, Philbrick WM (1994) Overexpression of parathyroid hormone-related protein in the skin of transgenic mice interferes with hair follicle development. Proc Natl Acad Sci USA 91:1133–1137

Yoshida H, Hayashi S, Kunisada T, Ogawa M, Nishikawa S, Okamura H, Sudo T, Shultz LD (1990) The murine mutation osteopetrosis is in the coding region of macrophage colony stimulating factor. Nature 345:442–444

Yoshizawa T, Handa Y, Uematsu Y, Takeda S, Sekine K, Yoshihara Y, Kawakami T, Arioka K, Sato H, Uchiyama Y, Masushige S, Fukamizu A, Matsumoto T, Kato S (1997) Mice lacking in the vitamin d receptor exhibit impaired bone formation, uterine hypoplasia and growth retardation after weaning. Nat Genet 16:391–396

7 Mechanotransduction: An Inevitable Process for Skeletal Maintenance

C.H. Turner, R.L. Duncan, and F.M. Pavalko

7.1 Bone Cells as Structural Engineers

The bony struts within cancellous regions of the skeleton follow trajectories that are nearly optimum for structural efficiency. This observation inspired Wolff to put forward his law of bone transformation, which can be paraphrased to state that trabecular trajectories align with the principal directions of mechanical stress and thus trabecular structures are optimized to their functional loads (Wolff 1892). In fact, trabecular patterns in bone often resemble structural designs derived from engineering computational algorithms that optimize structures by minimizing weight and maximizing stiffness (Fig. 1). The optimization of bones architecture occurs at every level of structural hierarchy, from the collagen fibrils in the matrix to the gross morphology. The collagen fibrils in long bones are laid down so that they provide the best tensile strength where it is needed (Portigliatti-Barbos et al. 1983; Takano et al. 1996), and the cross-sectional shape of long bones tends to provide the best

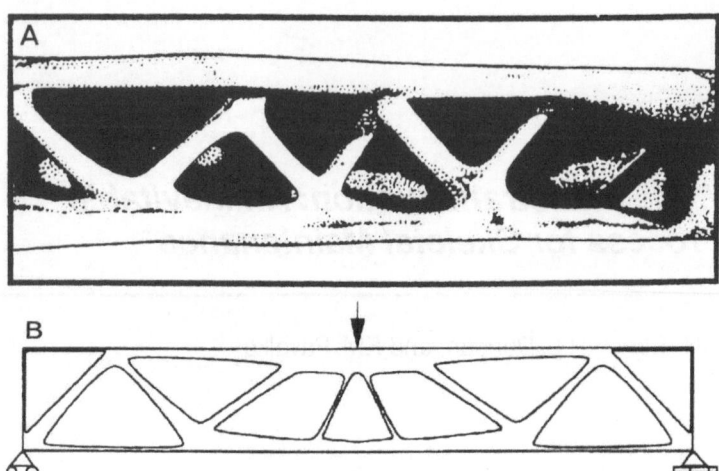

Fig. 1A, B. From an engineering viewpoint, skeletal structures appear to be optimized to their loading environment. The vulture wing bone (**A**), which typically withstands bending and shear stresses, has a trabecular strut pattern that resembles the structure of beam (**B**) optimized to minimize mass and maximize stiffness under bending loads, using a computer-aided algorithm. (From Bendsoe 1995)

resistance to bending stresses in the direction where bending is greatest (Pauwels 1980). Clearly, bone cells operate as if they were engineers, excavating and adding bone tissue to produce bony structures that are both complex and efficient, but from where do bone cells derive their engineering knowledge?

The structural blueprint for the skeleton is contained within the bone cells genetic program. When a young animal's limb is paralyzed, it continues to grow to nearly normal size and shape, even in the absence of mechanical loading (Lanyon 1980). Yet the paralyzed limb suffers from the lack of structural adaptation to mechanical loading; the bones in the limb have less mineral and their cross-sectional shapes are not mechanically efficient (Fig. 2). There is thus a component of skeletal design that is continuously updated by the mechanical environment of the organism. Bone cells begin with the genetic blueprint and sculpt it until the skeletal design meets the loading requirements. This process,

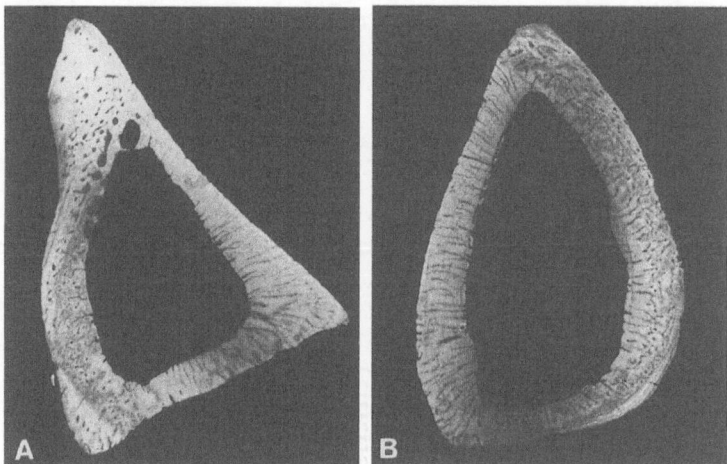

Fig. 2A, B. The bone in a paralyzed limb grows to a similar size and shape as a normal limb. However, there are subtle differences due to the lack of mechanical loading. For instance, paralysis causes the limb bones to have less mineral. Also paralysis inhibits the sculpting of the bone that occurs due to mechanical influences. This is illustrated in rats that were paralyzed in one hindlimb during the period of rapid skeletal growth: the cross-sectional shape of a normal tibia (**A**) is considerably different than that of a tibia of a paralyzed limb (**B**). (From Lanyon 1980)

termed *bone adaptation*, requires bone cells to detect mechanical signals in situ and integrate these signals into appropriate changes in the bone architecture.

7.2 How Does Bone Adaptation Work?

The most useful theory of bone adaptation is the "mechanostat" hypothesis put forth by Frost (1987). This theory proposes that there are a series of mechanical loading "thresholds" that determine when bone cells will be activated to remove or add new bone to the skeleton. These thresholds outline "windows" of mechanical usage (Fig. 3) and describe how windows of mechanical usage can be considered normal or pathological. When local mechanical signals in bone exceed the upper bound-

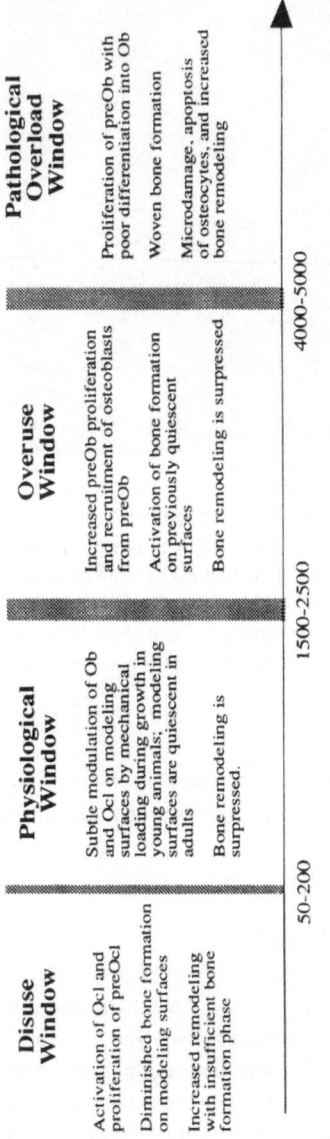

Disuse Window	Physiological Window	Overuse Window	Pathological Overload Window
Activation of Ocl and proliferation of preOcl	Subtle modulation of Ob and Ocl on modeling surfaces by mechanical loading during growth in young animals; modeling surfaces are quiescent in adults	Increased preOb proliferation and recruitment of osteoblasts from preOb	Proliferation of preOb with poor differentiation into Ob
Diminished bone formation on modeling surfaces		Activation of bone formation on previously quiescent surfaces	Woven bone formation
Increased remodeling with insufficient bone formation phase	Bone remodeling is surpressed.	Bone remodeling is surpressed	Microdamage, apoptosis of osteocytes, and increased bone remodeling
50–200	1500–2500	4000–5000	

Strain Stimulus (microstrain)

Fig. 3. Frost proposed that bone physiology is controlled mainly by mechanical influences and that bone adaptation can be grouped into four categories or "windows" according to the amount of mechanical loading. Disuse leads to increased bone remodeling and bone loss, while overuse leads to decreased bone remodeling and bone gain, and extreme overloading causes damage to the bone requiring increased bone remodeling for repair. Normal mechanical loading (physiological window) tends to repress bone remodeling and maintain bone mass. *Ob*, osteoblast; *Ocl*, osteoclast. (Frost HM, private communication)

ary of the physiological window, called the minimum effective strain (MES), bone will undergo modeling, or sculpting, and change its structure to reduce the local strains to below the MES. If the mechanical loads on the skeleton are very large, the bone strains will be pushed into a pathological overload zone causing woven bone formation on bone surfaces. Likewise there is a lower MES threshold below which bone tissue will be resorbed until the local strains are increased. Frost further suggested that certain hormones and biochemical agents may alter the mechanostat setting of bone to change the boundaries of the physiological window, allowing normal mechanical usage to increase bone mass and bone strength significantly.

The mechanisms by which the mechanostat works are unknown, however they require some form of cellular *mechanotransduction.* Mechanotransduction, or the conversion of a biophysical force into a cellular response, is an essential mechanism for a wide variety of physiologic functions which allow living organisms to respond to the mechanical environment. Presumably, mechanotransduction in bone must include four distinct phases: (1) *mechanocoupling*, the transduction of mechanical force applied to the bone into a local mechanical signal perceived by a sensor cell; (2) *biochemical coupling*, the transduction of a local mechanical signal into a biochemical signal and, ultimately, gene expression or protein activation; (3) *transmission of signal* from the sensor cell to the effector cell, i.e., the cell that will actually form or remove bone; (4) the *effector cell response,* the final tissue-level response.

A considerable volume of evidence has been gathered concerning each of the steps of mechanotransduction in bone. For instance, it now appears that bone cells may be more sensitive to the rate of applied strain rather than the peak strain magnitude (Fig. 4). The cells that detect mechanical signals are probably the osteocytes and bone lining cells (Fig. 5) and cell-to-cell signaling after a mechanical stimulus involves prostaglandins, especially those produced by the inducible isoform of cyclooxygenase (Chow et al. 1994; Forwood 1996), and nitric oxide (Fox et al. 1996; Turner et al. 1996). Finally, the bone formation response to mechanical loads involves both nonproliferating and proliferating osteoprogenitor cell populations. Mechanical loading induces an early bone formation response within 48 h that involves osteoblasts recruited from bone lining cells or nondividing preosteoblasts. Prolifer-

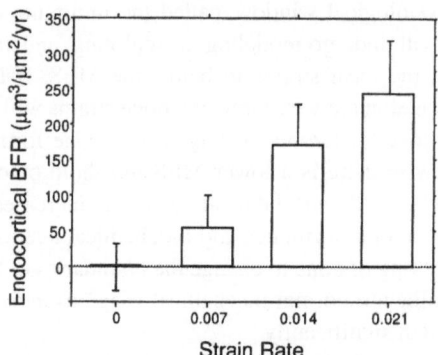

Fig. 4. A growing body of evidence suggests that the rate of applied strain, rather than peak strain, on the bone is a major determining factor for bone adaptation. In an experiment in which peak strains were held constant and strain rates were varied in the rat tibia, bone formation was proportional to applied strain rate (strain rate values were miscalculated in the original publication, and the corrected values are presented here). (Adapted from Turner et al. 1995a)

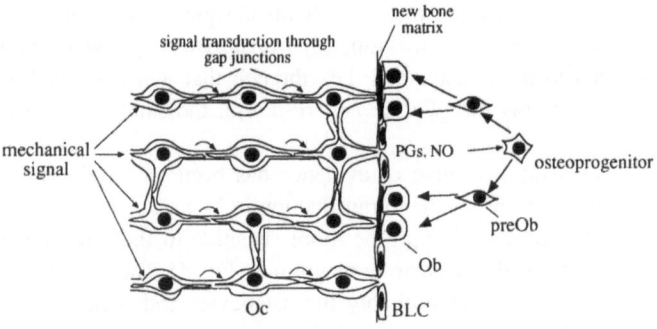

Fig. 5. It is hypothesized that osteocytes (*Oc*) and bone lining cells (*BLC*) detect mechanical signals and communicate those signals to the bone surface. Soluble mediators, which include prostaglandins (PGs) and nitric oxide (*NO*), are released and cause the recruitment and/or differentiation of osteoblasts (*Ob*) from proliferating and nonproliferating osteoprogenitor cells

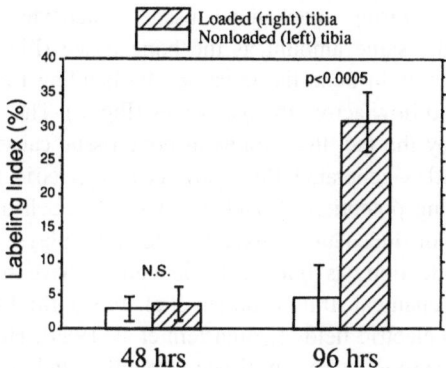

Fig. 6. Using a sustained release bromodeoxyuridine (BrdUrd) preparation, it has been shown that new osteoblasts from proliferating osteoprogenitor cells reach the bone surface about 72 h after a mechanical stimulus. BrdUrd labels cells as they undergo S phase, and thus marks proliferating cells. Some 48 h after mechanical loading in the rat tibia, new osteoblasts are present on the bone surface but are not labeled with BrdUrd indicating that they originated from nonproliferating precursors or bone lining cells; 96 h after mechanical loading, the number of osteoblasts on the bone surface remains elevated, but now 30%–40% of the new osteoblasts are labeled with BrdUrd. Therefore, mechanical loading can cause two distinct osteoblastic responses: an immediate response within 48 h in which osteoblasts are recruited from nondividing preosteoblasts and/or bone lining cells, and a delayed response involving proliferation and differentiation of preosteoblasts that requires three or more days. (Adapted from Turner et al. 1997)

ating osteoprogenitor cells are also stimulated by mechanical loading. These cells differentiate into osteoblasts 72–96 h after a mechanical stimulus (Fig. 6).

7.3 How Do Bone Cells Sense Mechanical Signals?

When loads are applied to bone, the tissue deforms causing local strains (typically reported in units of microstrain; 10 000 microstrain=1% change in length). We may assume that osteocytes act as the sensors of local bone strains because they are appropriately located in the bone for this function. During mechanical usage, osteocytes are probably stretched to the same amount as the bone tissue (Fig. 7). Also, the pressure gradients in bone tissue caused by bending forces create extracellular fluid flow across the osteocytes (Fig. 8). The issue is further complicated by the fact that strains in bone tissue cause piezoelectric effects and stress-generated fluid flow causes electric fields in bone called streaming potentials (Chakkalakal 1989). Each of these tissue-level effects of mechanical loading probably plays some role in mechanotransduction, as bone cells in culture have been shown to respond to mechanical strain (Somjen et al. 1980), fluid flow (Reich et al. 1990), and electric fields (Korenstein et al. 1984). However, it now appears that the most important tissue-level effect in bone may be fluid flow. Since bone is a stiff, brittle material, mechanical strains in the tissue tend to be small and the maximum strains on osteocytes in vivo (probably less than 5000 microstrain) are considerably less than the

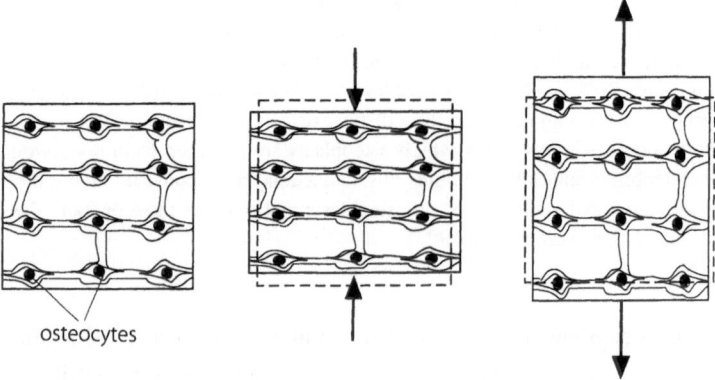

osteocytes

Fig. 7. Osteocytes within the bone matrix are thought to be the sensors of mechanical signals. When a region of bone is compressed or stretched, it expands or contracts in the perpendicular direction. This phenomenon is called Poissons effect and causes a biaxial strain field on osteocytes embedded in the matrix. However the deformation of osteocytes in vivo is probably limited to less than 5000 microstrain due to the brittleness of the bone mineral (bone tissue begins to fail at about 7000 microstrain). In cell culture, strains typically must exceed 10 000 microstrain to cause an anabolic response in bone cells. (From Duncan and Turner 1995)

strains necessary to activate bone cells in vitro (10 000 microstrain). Furthermore, bone mechanotransduction in vivo is highly dependent upon strain rate rather than strain magnitude (Fig. 4), which suggests that some dynamic signal, like fluid flow, activates the mechanotransduction cascade. Fluid flow in bone causes fluid shear stresses on osteocytes and electric potentials due to streaming effects. Each of these signals, electrical and mechanical, may play a role in mechanotransduction, although cell culture experiments suggest that the fluid shear effects might be more effective than the electrical fields for stimulating osteoblasts (Hung et al. 1996a; Reich et al. 1990).

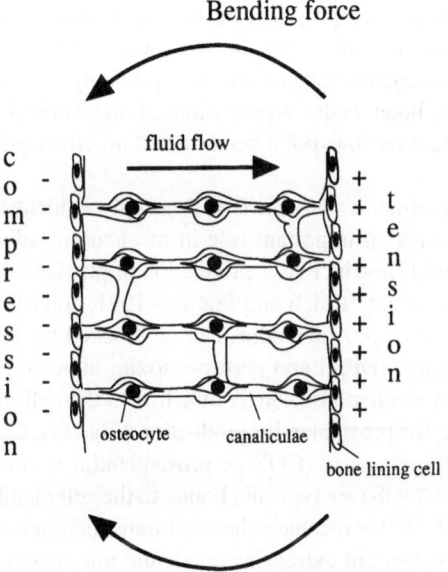

Fig. 8. Bending forces not only cause deformation of osteocytes, but generate pressure gradients that drive fluid flow through the canalicular spaces. Bending causes compressive stress on one side of the bone and tensile stresses on the other. This leads to a pressure gradient in the interstitial fluids that drives fluid flow from regions of compression to tension. Fluid flows through the canaliculae and across the osteocytes providing nutrients and causing flow-related shear stresses on the cell membranes. The fluid flow also creates an electric potential called a streaming potential. (From Duncan and Turner 1995)

7.4 How Do Bone Cells Transduce a Mechanical Signal?

There is now evidence of several mechanochemical transduction pathways within bone cells. The results reported here were generated using two different systems to stimulate bone cells mechanically in culture. Fluid shear stress was applied to bone cells cultured in monolayer using a fluid flow loop (Fig. 9). Mechanical strain was applied to bone cell culture using four-point bending (Fig. 10). The four-point bending apparatus also allowed us to vary the fluid forces on the cells by varying the displacement rate of the culture plate; the fluid pressure and fluid shear stress on the cultured cells is proportional to the displacement rate. Osteoblastic expression of c-*fos*, cyclooxygenase (COX)-2, and osteopontin genes was proportional to displacement rate (Owan et al. 1997; Fig. 11), indicating that these genes, which are important for osteoblastic mechanotransduction, are probably regulated by fluid forces on the bone cells. Application of mechanical strains up to 5600 microstrain by four-point bending had no effect osteoblastic gene expression.

As our experiments show, it is increasingly evident that fluid flow within bone plays an important role in mechanotransduction. In bone cell culture, fluid flow increases production of prostaglandins and nitric oxide within minutes (Reich and Frangos 1991; Johnston et al. 1996). This stimulation of prostaglandins can be blocked by 70%–80% by G protein inhibitors GDPβS and pertussis toxin, indicating that a G protein-associated mechanotransducer attached to the cell membrane may be responsible for prostaglandin production (Fig. 12). Constitutive isoforms of cyclooxygenase (COX or prostaglandin synthase) and nitric oxide synthase (NOS) are typically bound to the cell membrane and thus would be available for mechanochemical transduction involving G proteins. The movement of extracellular calcium ions across the cell membrane also appears to play a role in mechanotransduction. Prostaglandin production was inhibited by almost 90% when extracellular calcium was chelated (Reich et al. 1997).

One way in which extracellular calcium can pass across the cell membrane is through a stretch-activated cation channel (Duncan et al. 1996). Hypotonic swelling of osteoblasts causes a rise in intracellular calcium within 1 min (Fig. 13). The addition of gadolinium chloride ($GdCl_3$), which blocks the stress-activated calcium channels in the cell

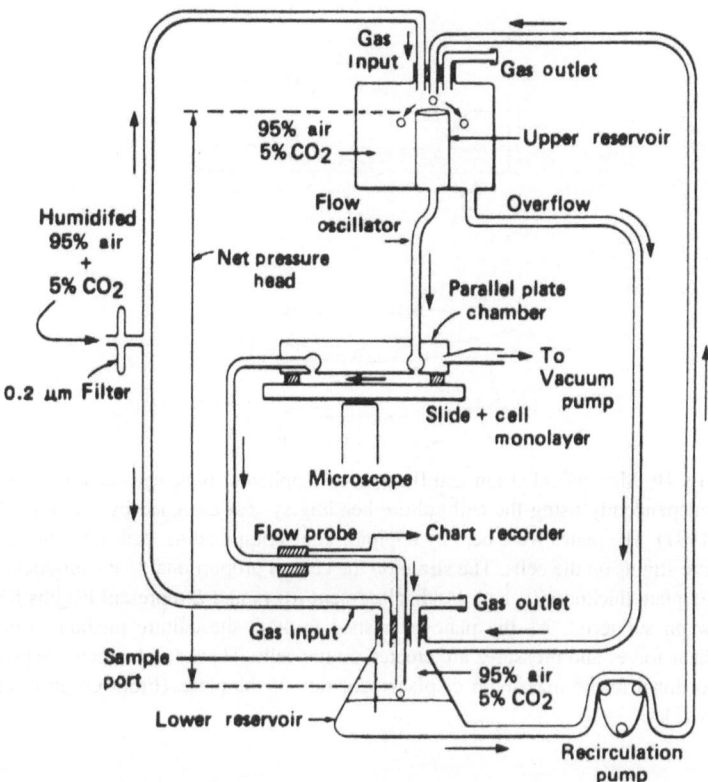

Fig. 9. The effects of fluid flow on bone cells can be studied using a flow loop configuration similar to that designed by Frangos et al. (1985). This system allows one to control the flow of culture medium across the cell monolayer, thus the fluid shear stress on the cells can be carefully controlled. (From Frangos et al. 1985)

membrane, eliminates the intracellular calcium response to hypotonic swelling. Likewise an increase in intracellular calcium occurs within minutes in osteoblasts exposed to fluid shear stresses, and this response is partially blocked by GdCl$_3$ (Hung et al. 1996b). Fluid shear stress also causes a release of calcium from intracellular stores. It is not clear yet what interaction the inositol triphosphate pathway, which controls re-

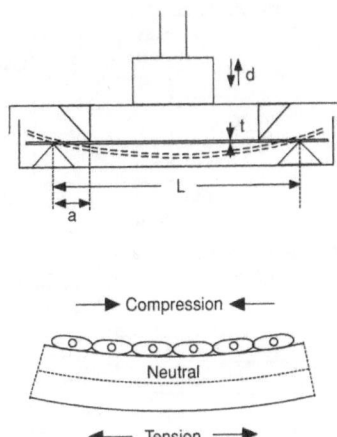

Fig. 10. Mechanical strain and fluid forces applied to bone cells can be varied independently using the cell culture bending system designed by Owan et al. (1997). The plates were bent by applying four-point loading, causing compressive strains on the cells. The strain on the cells is proportional to the product of the plate thickness (t) and the displacement (d) (a and L represent lengths between supports). As the plate is pushed through the culture medium, fluid shear forces and pressures are created on the cells. These fluid effects are proportional to the maximum displacement rate of the plate. (From Owan et al. 1997)

lease of intracellular calcium, has with the stretch-activated cation channel or the G protein mechanotransducer, but these systems may be interrelated.

Another pathway by which mechanotransduction occurs is through the integrin-cytoskeleton complex. Integrins are heterodimeric transmembrane proteins that bind to the extracellular matrix (ECM) on the outside of cells and are linked to the actin cytoskeleton via the short cytoplasmic domain of the β subunit on the inside of cells as specialized sites known as focal adhesions (Fig. 14). Several lines of evidence using various cell types, including fibroblasts, epithelial cells, endothelial cells, neutrophils, and osteoblasts, indicate that a key molecule in mediating linkage of actin filaments to integrin cytoplasmic domains is the protein α-actinin (Otey et al. 1990; Pavalko and Burridge 1991; Pavalko

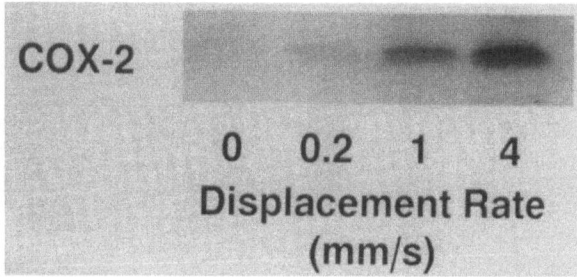

Fig. 11. Using the four-point cell culture bending system (see Fig. 10), one can show that fluid forces are more important than mechanical strain for inducing gene expression in osteoblasts (MC3T3-E1 cells). In the experiment shown, inducible cyclooxygenase (*COX-2*) gene expression 1 h after mechanical loading is clearly dependent upon the displacement rate of the culture plate. This result was anticipated because fluid forces on the cultured cells are proportional to the displacement rate. It should be noted that the maximum strain on the bone cells (5600 microstrain) was held constant in this experiment, so gene expression did not appear to be dependent upon strain magnitude

and LaRoche 1993). Fluid flow on bone cells induces recruitment of integrins to focal adhesions and causes the actin filaments in the cell to reorganize into large bundles of actin filaments called stress fibers (Fig. 15). Coincident with fluid flow-induced changes in cytoskeletal organization is an increase in expression of (COX-2), an enzyme important in bone mechanotransduction (Forwood 1996). Our experiments indicate that microinjection into osteoblasts of a 53 kDa proteolytic fragment of α-actinin, which contains the integrin binding domain, but not the actin binding domain, causes the competitive displacement of the cells endogenous α-actinin from focal adhesions (Pavalko and Burridge 1991) and blocks the fluid flow-induced development of actin stress fibers in cells (Fig. 16B). In addition, microinjection of this fragment also prevented the induction of COX-2 which occurs normally in osteoblasts within 1 h after exposure to fluid flow (Fig. 16D). These results demonstrate that anchorage of actin filaments at focal adhesions via α-actinin is necessary for both the formation of stress fibers and up-regulation of COX-2 expression that is normally induced when osteoblasts are subjected to fluid shear. Thus, blocking the formation of actin stress fibers somehow prevents transmission of a signal to the

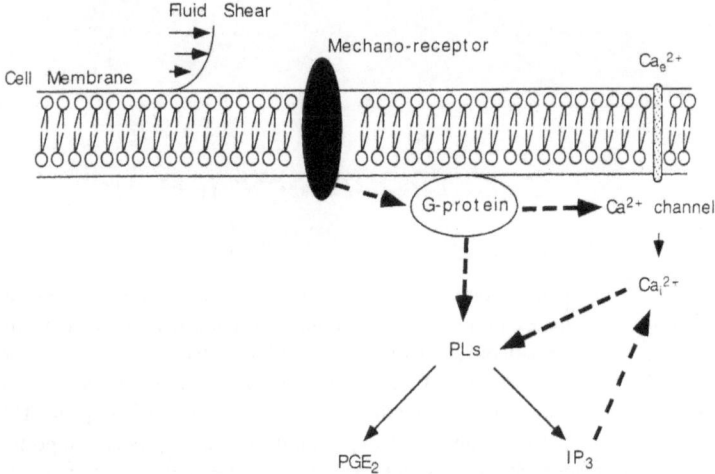

Fig. 12. A G protein-associated mechanochemical signal transducer mediates the activity of constitutive isoforms of the cyclooxygenase and nitric oxide synthase enzymes. These enzymes are activated in osteoblasts and osteocytes within minutes after a mechanical stimulus and cause the production of prostaglandins and nitric oxide. Inhibitors of G proteins cause over 80% reduction in the production of prostaglandins. Calcium channels in the cell membrane also appear to be involved in the signal pathway. (From Reich et al. 1997)

nucleus that is required for expression of COX-2 in response to fluid shear.

In summary, two putative mechanochemical pathways have been identified in bone cells. One involves a mechanotransducer that resides in the cell membrane. This transducer is G protein-linked and also interacts with a stretch-activated cation channel, the inositol triphosphate messenger pathway, and constitutive isoforms of COX and NOS. The second pathway involves a direct linkage between the transmembrane integrins, the actin cytoskeleton, and the nuclear transcription machinery. The induction of COX is dependent upon this pathway. Interesting, inducible COX was not observed in the cell membrane (where constitutive COX is found) but was seen in the nuclear membrane and in the cell nucleus (Fig. 16D).

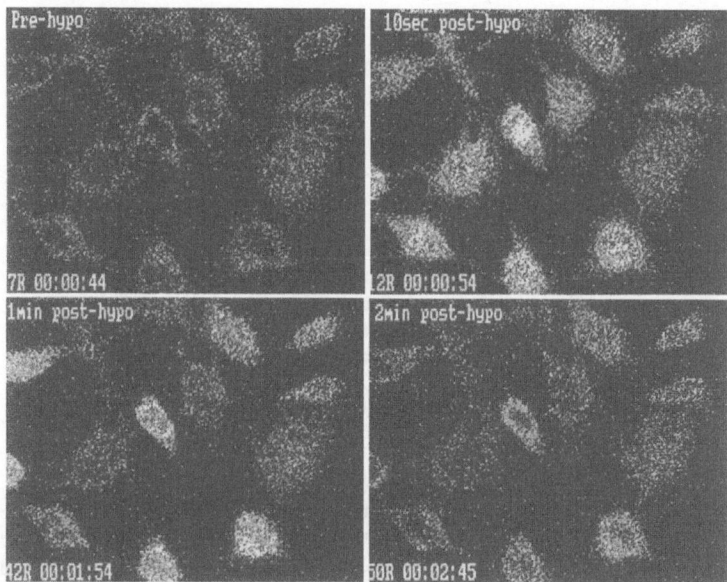

Fig. 13. Intracellular calcium (*white specks*) in MC3T3-E1 osteoblasts after exposure to hypotonic swelling. Adding gadolinium chloride ($GdCl_3$), which blocks the stress-activated calcium channels in the cell membrane, completely blocks the intracellular calcium response. Thus stretch-activated membrane channels are important mechanochemical transducers for the intracellular calcium messenger system

7.5 The Effects of Aging and Hormones on Mechanotransduction

In Frost's theory, hormones and aging (for example) modulate bone physiology by adjusting the mechanostat. Aging and at least one hormone appear to fit his theory.

Mechanically induced bone formation in the rat tibia was decreased 16-fold in 19 month-old rats compared to 9 month-olds (Fig. 17). It is unclear why aging inhibits mechanotransduction, but its effects are probably multiple. Possible sensors of mechanical signals include bone lining cells and osteocytes and the populations of these cell types de-

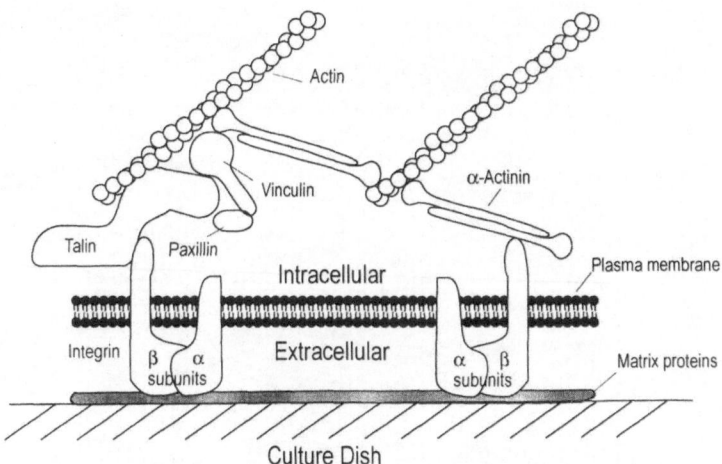

Culture Dish

Fig. 14. The cytoskeletal components at the point of attachment with the extracellular matrix in vitro (adapted from Pavalko et al. 1991a). The integrin, made
up of two heterodimers α and β, spans the plasma membrane of the cell. The
extracellular domain binds to extracellular matrix proteins. The intracellular
domain interacts with either talin or α-actinin (Pavalko and Otey 1994). Either
of these proteins, in turn, attach to actin. Vinculin and paxillin also may have a
role in local adhesions. Other proteins, e.g., tensin, have been identified in focal adhesions but are not shown here

crease with advancing age. Futhermore, the efficiency of osteoblasts to
respond to mechanical signals may be compromised with age.

Parathyroid hormone (PTH) is one hormone that appears to interact
directly with cellular mechanotransduction. Mechanical loading fails to
induce bone formation in thyroparathyroidectomized rats, but when
these rats are given a dose of PTH shortly before mechanical loading,
the mechanically induced bone formation response is restored (Chow et
al. 1997). This result demonstrates an interaction between hormonal and
mechanical signals and supports to some extent the prospectus put forth
by Frost, that is, cellular responsiveness to mechanical signals is a
fundamental aspect of bone biology; hormones and other systemic factors act at least in part by modulating cellular mechanotransduction
pathways.

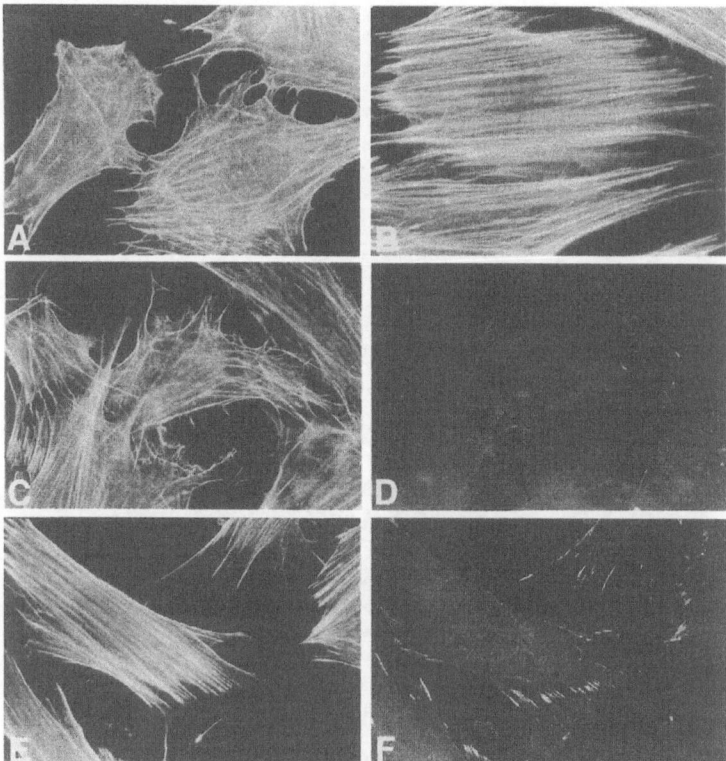

Fig. 15A–F. MC3T3-E1 osteoblasts subjected to fluid shear (12 dynes/cm^2) for 60 min undergo dramatic reorganization of the actin cytoskeleton. **A** Control cells not subjected to flow have poorly organized stress fibers labeled with Texas red-phalloidin. **B** Cells subjected to fluid flow for 60 min develop prominent stress fibers labeled with Texas red-phalloidin that are aligned roughly parallel to each other. **C, D** Control cells not subjected to fluid shear which have poorly organized stress fibers (**C**) also do not concentrate β1 integrins into focal adhesions (**D**). **E, F** Cells subjected to fluid flow for 60 min not only develop prominent stress fibers labeled with Texas red-phalloidin (**E**), but also recruited β1 integrins into focal adhesions (**F**)

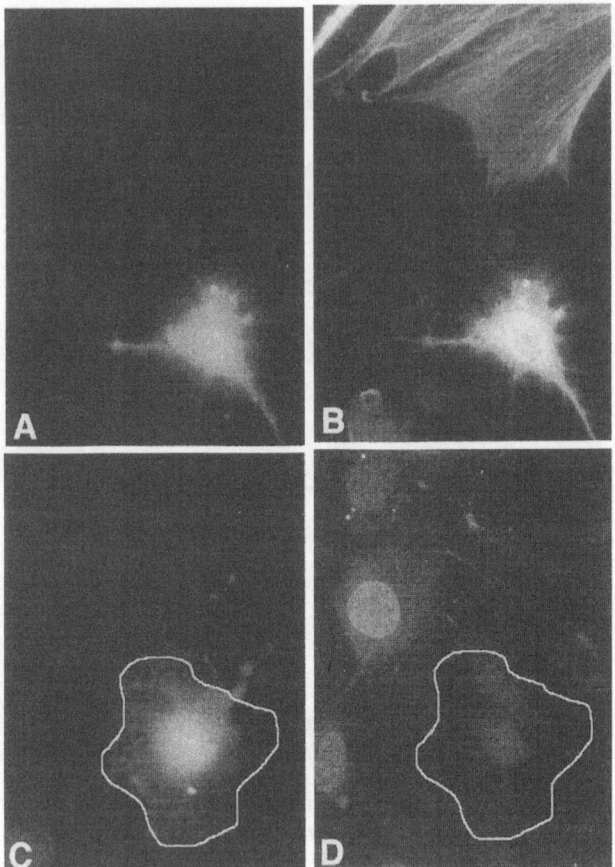

Fig. 16 A–D. Injection of a protolytic α-actinin fragment into osteoblasts inhibited stress fiber formation. Cells were microinjected with the 53 kDa α-actinin fragment which was conjugated directly to rhodamine (injected cells are identified by the rhodamine fluorescence seen in **A** and **C**). This fragment replaces α-actinin in the cell but does not form a link between the integrins and the f-actin fibers (Pavalko et al. 1991b). Cells were returned to the incubator for 1 h, then subjected to fluid shear at 12 dynes/cm^2 for 1 h before being fixed and stained for stress fibers by labeling with FITC-phalloidin (**B**), or for cyclooxygenase (COX-2) (**D**) by labeling with a specific antibodies. Compared to their uninjected neighbors, injected cells fail to develop stress fibers (**B**), and do not up-regulate COX-2 expression (**D**; the position of the injected cells in panel panels **C** and **D** is shown by the outline). This experiment demonstrates the importance of stress-fiber formation for the production of COX-2

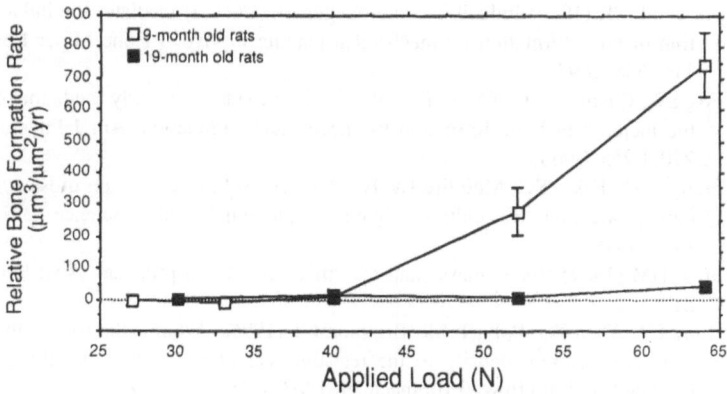

Fig. 17. Relative bone formation rate on the endocortical surface of the tibia for 9 month-old and 19 month-old rats subjected to four-point bending. For applied loads of 52 and 64 N, mechanically induced bone formation rate was over 16-fold higher in younger rats than in the old rats. (From Turner et al. 1995b, with permission from the American Society for Bone and Mineral Research)

References

Bendsoe MP (1995) Optimization of structural topology, shape, and material. Springer-Verlag, Berlin, Heidelberg, New York

Chakkalakal DA (1989) Mechanoelectric transduction in bone. J Mater Res 4: 1034–1046

Chow JW, Chambers TJ (1994) Indomethacin has distinct early and late actions on bone formation induced by mechanical stimulation. Am J Physiol 267: E287–E292

Chow JW, Fox S, Jagger CJ, Chambers TJ (1997) A role for parathyroid hormone in the mechanical responsiveness of rat bone. J Bone Miner Res 12: S316

Duncan RL, Turner CH (1995) Mechanotransduction and the functional response of bone to mechanical strain. Calcif Tissue Int 57: 344–358

Duncan RL, Kizer N, Barry EL, Friedman PA, Hruska KA (1996) Antisense oligodeoxynucleotide inhibition of a swelling-activated cation channel in osteoblast-like osteosarcoma cells. Proc Nat Acad Sci USA 93: 1864–1869

Forwood MR (1996) Inducible cyclo-oxygenase (COX-2) mediates the induction of bone formation by mechanical loading in vivo. J Bone Miner Res 11: 1688–1693

Fox SW, Chambers TJ, Chow JW (1996) Nitric oxide is an early mediator of the increase in bone formation by mechanical stimulation. Am J Physiol 270: E955–E960

Frangos JA, Eskin SG, McIntire LV, Ives CL (1985) Flow effects on prostacyclin production by cultured human endothelial cells. Science 227: 1477–1479

Frost HM (1987) Bone "mass" and the "mechanostat": a proposal. Anat Rec 219: 1–9

Hung CT, Allen FD, Pollack SR, Brighton CT (1996a) What is the role of the convective current density in the real-time calcium response of cultured bone cells to fluid flow? J Biomech 29: 1403–1409

Hung CT, Allen FD, Pollack SR, Brighton CT (1996b) Intracellular Ca2+ stores and extracellular Ca2+ are required in the real-time Ca2+ response of bone cells experiencing fluid flow. J Biomech 29: 1411–1417

Johnson DL, McAllister TN, Frangos JA (1996) Fluid flow stimulates rapid and continuous release of nitric oxide in osteoblasts. Am J Physiol 271: E205–E208

Korenstein R, Somjen D, Fischler H, Binderman I (1984) Capacitative pulsed electric stimulation of bone cells. Induction of cyclic-AMP changes and DNA synthesis. Biochim Biophys Acta 803: 302–307

Lanyon LE (1980) The influence of function on the development of bone curvature. An experimental study on the rat tibia. J Zool Lond 192: 457–466

Owan I, Burr DB, Turner CH, Qiu J, Tu Y, Onyia JE, Duncan RL (1997) Mechanotransduction in bone: osteoblasts are more responsive to fluid forces than mechanical strain. Am J Physiol 273: C810–C815

Otey CA, Pavalko FM, Burridge K (1990) An interaction between α-actinin and the b1 integrin subunit in vitro. J Cell Biol 111:721–730

Pauwels F (1980) Biomechanics of the locomotor apparatus. Springer-Verlag, Berlin, Heidelberg, New York

Pavalko FM, Otey CA, Simon KO, Burridge K (1991a) α-Actinin: a direct link between actin and integrins. Biochem Soc Trans 19: 1065–1069

Pavalko FM, Burridge K (1991) Disruption of the actin cytoskeleton after microinjection of proteolytic fragments of alpha-actinin. J Cell Biol 114: 481–491

Pavalko FM, LaRoche SM (1993) Activation of human neutrophils induces an interaction between the integrin β2 subunit (CD18) and the actin-binding protein α-actinin. J Immunol 151:3795–3807

Pavalko FM, Otey CA (1994) Role of adhesion molecule cytoplasmic domains in mediating interactions with the cytoskeleton. Proc Soc Exp Biol Med 205:282–293

Portigliatti-Barbos M, Bianco P, Ascenzi A (1983) Distribution of osteonic and interstitial components in the human femoral shaft with reference to structure, calcification, and mechanical properties. Acta Anat 115: 178–186

Reich KM, Gay CV, Frangos JA (1990) Fluid shear stress as a mediator of osteoblast cyclic adenosine monophosphate production. J Cell Physiol 143: 100–104

Reich KM, Frangos JA (1991) Effect of flow on prostaglandin E_2 and inositol triphosphate levels in osteoblasts. Am J Physiol 261: C428–C432

Reich KM, McAllister TN, Gudi S, Frangos JA (1997) Activation of G proteins mediates flow-induced prostaglandin E2 production in osteoblasts. Endocrinology 138: 1014–1018

Somjen D, Binderman I, Berger E, Harell A (1980) Bone remodeling induced by physical stress is prostaglandin E_2 mediated. Biochim Biophys Acta 627: 91–100

Takano Y, Turner CH, Burr DB (1996) Elastic anisotropy of osteonal bone is dependent on the mechanical strain distribution. J Bone Miner Res 11: S268

Turner CH, Owan I, Alvey T, Hulman J, Hock JM (1997) Recruitment and proliferative responses of osteoblasts after mechanical loading in vivo determined using sustained-release bromodeoxyuridine. Bone 22: 463–469

Turner CH, Takano Y, Owan I, Murrell GAC (1996) Nitric oxide inhibitor L-NAME suppresses mechanically-induced bone formation in rats. Am J Physiol 270: E634–E639

Turner CH, Owan I, Takano, Y (1995a) Mechanotransduction in bone: role of strain rate. Am J Physiol 269: E438–E442

Turner CH, Takano Y, Owan I (1995b) Aging changes mechanical loading thresholds for bone formation in rats. J Bone Miner Res 10: 1544–1549

Wolff J (1892) Das Gesetz der Transformation der Knochen. Hirschwald, Berlin

8 The Regulation of Gene Expression in Bone by Mechanical Loading

T.M. Skerry

8.1 Introduction

In the search for therapies to increase bone mass, it is not surprising that much attention has focused on studies of the effects of mechanical loading on the skeleton. Historical comments on the stronger skeletons of individuals with active lifestyles are attributed to Galileo, but serious studies of the relationships between bone mass and loading were not made until the late 1800s, when Culmann, von Meyer, Roux and Wolff initiated what has now become the idea of functional adaptation in the skeleton. Since then, studies have been predominantly phenomenological, cataloguing the changes in bone mass in response to exercise, disuse and applied loading. That is not to say that the data have not had important applications. For example, studies which have determined the numbers, rates and magnitudes of loads which influence bone mass have been useful in order to develop appropriate exercise regimens for improvement of bone mass. In addition it has become clear from clinical studies that exercise which fails to increase bone mass may still reduce the risk of fractures in osteoporotic individuals. This is because the increased muscle power and coordination which accompany increased fitness reduce the risk of falling, and therefore incidence of fracture.

In more recent years though, the dramatic advances in molecular biological techniques have led to the development of more mechanistic approaches to the study of mechanical influences on bone cells both in vivo and in vitro. The results of such experiments are still at early stages of development. However, in the medium and long term, novel therapeutics may be developed. These could be targeted to the process downstream of the transduction of a mechanical event, thereby impacting on the biochemical processes within a bone cell. This chapter will review some of the studies from the past and present the results of our approach to discovery of novel targets using differential display techniques to identify genes regulated in bone by mechanical loading.

8.2 Rationale for Studies on Loading of the Skeleton

Exercise has a number of direct and indirect effects on the occurrence of osteoporotic fractures. Even the most mild exercise increases cardiovascular performance, and general feelings of well-being in people of all ages, which may in turn motivate the individual to perform more active pursuits which have direct benefits on skeletal strength. Higher levels of activity may contribute to direct loading-related stimulation of enhanced bone formation (Smith and Raab 1986), although in the elderly, it is rare to achieve sufficiently intense stimulation of the skeleton by loading that such effects are statistically significant in populations (Bassey and Ramsdale 1995). However even in the absence of significant exercise mediated effects on bone mass, activity will reduce incidence of fractures in osteoporotic individuals because of the increased muscle strength and co-ordination which invariably accompany regular exercise. This is because fractures in osteoporotics are almost invariably the result of falling, and increased co-ordination and strength impact clearly on balance and therefore ability to resist falls (Smith and Gilligan 1991). Finally, mechanical loading of bone is a suitable subject for research in order to determine the cellular mechanisms of mechanotransduction as therapeutic targets. All of these goals require an understanding of some of the important characteristic features of mechanical loading regimens which must be considered in either clinically or physiologically relevant model systems designed to mimic exercise which builds new bone.

8.3 Physiologically Relevant Strain Regimens

The way in which the effect of loading is measured is to describe the strain in the tissue. Strain is defined by the relationship between the dimensions of an object before and during application of a load. The alteration in size divided by the original size is the strain, which as a ratio, is a unitless number (Fig. 1). Bone strains are usually quoted as microstrain - strain 10^{-6} (μE). Compressive strains are usually given a negative sign, while tensile strains are positive.

Typical peak bone strains are usually in the range of 2000–3000 μE (Rubin and Lanyon 1984). It is often considered that strains in the

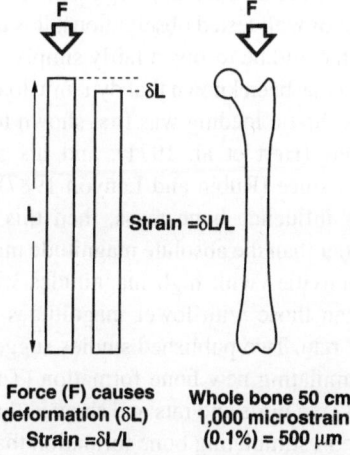

Force (F) causes   Whole bone 50 cm
deformation (δL)   1,000 microstrain
Strain =$\delta L/L$ (0.1%) = 500 μm

Fig. 1. Strain is the ratio of deformation divided by original length, so as a ratio it has no units. Typical long bone strains are in the range of 1000–3000 10^{-6}

higher zone of that range stimulate bone formation, while lower ones do not (Turner et al. 1995b). This ignores some fundamental principles regarding the differences between different bones in the skeleton (Hylander et al. 1987; Hylander and Johnson 1997). Bones vary in the strains that they experience and the most extreme ends of that spectrum are probably the long bones, for example the tibia, and the skull (Hillam et al. 1994). It is interesting to see that, at those two sites, the peak strains engendered by even extremes of activity differ by an order of magnitude, so that the skull experiences strains of 200 μE while the tibia experiences over 2000 (Tables 1, 2). Such a result suggests that it is not only the magnitude of strain at a given site which governs the response of the bone to load, but some positionally derived information on the nature of strains to which the bone in question is accustomed.

So while it is clear that strain magnitude is one of the most important determinants for the effectiveness of an exercise regimen to affect bone mass, several others are also important (Skerry 1997). With the exception of studies on strain magnitude, the effectiveness of different components of mechanical loading regimens in stimulating bone mass is a subject which has received relatively little attention. Because of this, it is hard to put forward completely proven arguments for the importance of individual components of a loading regimen. However, by building on the small number of well tested observations, it is possible to identify several issues which could be resolved fairly simply

For many years it has been known that dynamic loads were necessary to affect bone mass. Static loading was first shown to be ineffective in this context by Hert (Hert et al. 1971), and his studies have been supported by others since (Rubin and Lanyon 1987). If cyclical loads but not static ones influence bone mass, then this suggests that the change in strain rather than the absolute magnitude may be a controlling factor. Naturally, activities with high magnitudes would induce more change in strain than those with lower magnitudes if both were performed at the same rate. Two published studies suggest the importance of strain rate in stimulating new bone formation (Turner et al. 1995a; O'Connor et al. 1982) in birds and rats and showed that high strain rates were more effective in stimulating bone formation than low ones. What is also interesting in this context is that the direction of high rate strains is less important than the rate. In a study to confirm the rate dependence of bone formation in response to mechanical loading in vivo using a rat

Table 1. Maximum strains recorded on a human subject

Activity	Maximum principal strain ($\mu\varepsilon$)		Shear strain ($\mu\varepsilon$)	
	Skull	Tibia	Skull	Tibia
Eating	130	–	240	–
Walking	50	350	60	1060
Heading ball	200	840	350	1250
Jump (0.45 m)	170	850	190	1200
Jump (1 m)	–	2060	–	2900

From Hillam (unpublished data).
In both cases the strains and rates in the skull are normally less than 10% of those in the tibia, suggesting some regional differences in requirement for strain at the two sites.

Table 2. Maximum strains recorded on a human subject

Activity	Maximum principal strain rate ($\mu\varepsilon/s$)		Maximum shear strain rate ($\mu\varepsilon/s$)	
	Skull	Tibia	Skull	Tibia
Eating	2 600	–	5 300	–
Walking	1 600	11 800	2 100	10 500
Heading ball	30 100	15 100	47 800	24 000
Jump (0.45 m)	44 000	180 800	51 600	244 200

From Hillam (unpublished data).
In both cases the strains and rates in the skull are normally less than 10% of those in the tibia, suggesting some regional differences in requirement for strain at the two sites.

model, we investigated the effect not only of slow and fast rates, but also of asymmetrical wave forms with fast rise/slow fall and slow rise/fast fall ramps (Skerry and Peet 1997). Interestingly, those studies showed that a fast fall ramp of loading is as effective as the same rate of rise with a slow fall (Fig. 2).

This result raises the possibility that rate and not magnitude is the important governing stimulus in the response of bone to loading. However, in the same study on human bone strain in the skull and tibia, we

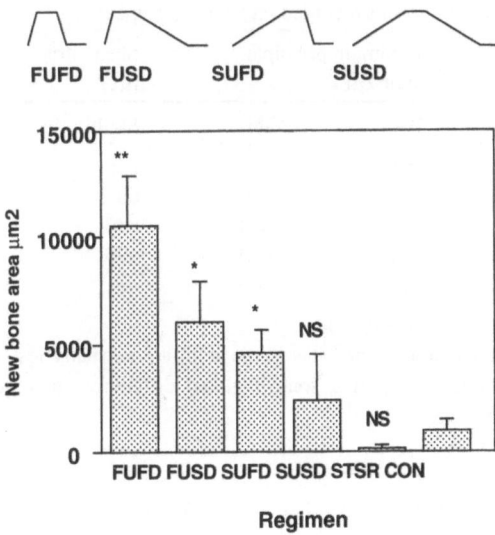

Fig. 2. Effects of different loading regimens on bone formation in vivo. The *upper panel* shows the fast up fast down (*FUFD*), fast up slow down (FUSD), slow up fast down (*SUFD*) and slow up slow down (*SFSD*) rate ramp wave forms applied to rat forelimbs in vivo for 300 cycles per day. Fast rates are 100 000 µE/s, and slow rates are 4000 µE/s. Static strains were slow up and down ramps applied for 300 s, and controls were not loaded. The histogram shows significant new bone formation was induced by FUFD (p.001) and by FUSD and SUFD (p0.01). The last two were indistinguishable in their response. STSR and control did not form significant amounts of new bone

showed that the rates of strain at the two sites showed similarly large differences (Fig. 1), suggesting that there is no absolute effective strain, but that an effective strain stimulus relates to some degree of overstrain above the customary level of exposure of each bone.

Naturally there are numerous other components of a strain regimen which affect its ability to influence bone remodeling. Among those are distribution, frequency and to some degree duration (Skerry 1997). Full consideration is outside the scope of this chapter, but it is worth noting that of those, duration may be the least important as there is evidence that only a very few loading cycles cause significant changes in bone mass in animals, while maximal osteogenesis can be induced by loading for about 1 min in each 24 h period in experimental model systems (Rubin and Lanyon 1987). It is dangerous to extrapolate too much the minimum duration of daily exercise needed in humans, but it is probable that less than 30 min of exercise per day confers the maximal benefits of exercise on human bone (Bassey and Ramsdale 1994).

8.4 Cells Responsive to Loading

It is probable that almost all bone cells are affected indirectly by the effects of mechanical loading. In nature there are in fact few cells which do not exhibit responsiveness to mechanical stimulation. However, the important feature of a load-related response in the context of a functionally adaptive response of the skeleton is to determine whether the effect is a nonspecific trauma response to excessive high strains which cause damage, or one due to a small deviation from physiological strain for cells at that skeletal site.

The ability to sense strains in bone requires close attachment of the sensing cells to the bone. It follows logically therefore that osteoblasts and osteocytes may be more involved in direct mechanosensing than other cell types. Opinions vary as to the importance of either osteocytes or osteoblasts cells in mechanosensing (Duncan and Turner 1995; Mullender and Huiskes 1997), as there is only circumstantial evidence for their role. Both cell types have been shown to be responsive to strain within short times of the start of osteogenic loading (Skerry et al. 1989). However, load responsiveness is not proof of involvement in mechanosensing. Although osteocytes are distributed throughout the matrix and communicate with each other (Doty 1981; Jones et al. 1993), so that they form an ideally located population of strain sensors, their origin as osteoblasts leaves the possibility that their load responsiveness is vestigial. The opposite hypothesis is also possible: that osteoblasts are responsive to strains because such a feature is necessary for their role as osteocytes. As with most biological questions, the likelihood is that neither of those extremes is correct, but that both cell types play a role in mechanosensing. The highly orchestrated influence of loading on periosteal, endosteal and intracortical remodeling makes unlikely the possibility that it is controlled simply by either osteocytes or bone surface cells. If that is true, then one essential feature in such a system is that osteocytes must be able to communicate somehow with surface cells. This is because their position precludes any direct action in either forming or resorbing bone matrix. Information derived from loading events can only be acted upon at the surface, where new bone can be deposited or osteoclasts can be recruited and stimulated to resorb bone (Vaes 1988).

While it is has been shown that loading or disuse can also alter marrow cell activity, both in vivo and in vitro (Thomas and ElHaj 1996; Keila et al. 1994), it is not certain that this is a direct effect relevant to functional adaptation. Pressure changes within the medullary cavity during loading could affect marrow cells directly, but it is unlikely that these are high (Wozasek et al. 1994). What seems more likely is that changes at the bone surface are involved in the formation, recruitment or attraction of marrow stromal cells to the sites where their action is needed.

8.5 Load-Induced Responses

In order to understand the responses of bone to loading, numerous studies have been performed both in vivo and in vitro to detail the changes induced in bone and bone cells in response to different forms of loading. The sequence of events which occur after loading is presented in Table 3. Both in vivo and in vitro methods have advantages and disadvantages, and the one which causes the largest degree of difficulty for interpretation of data is that in vitro models lack the ability to relate loading-induced changes to alterations in bone mass. Against this, in vivo models rarely allow the specific dissection of responses that are possible with in vitro systems. In order to address those questions we have used a combination of in vivo/in vitro methods. These techniques have an advantage over culture methods used by others as we can study the alterations in gene expression (for example) which precede load-induced bone formation if the same stimulus is applied repeatedly in vivo.

The in vivo rat ulna loading model we use was devised by Torrance et al. (1994) (Fig. 3). This involves the application of compressive forces between the carpus and elbow of a rat which caused the already curved bone to bend more. In a series of studies, this model has been used to

Table 3. Events following bone loading in vivo and in vitro

Time after loading	Observed effect on cells	Reference
100 ms	↑ Intracellular calcium	Jones et al. (1995)
5 s–15 min	↑ Phospholipase A$_2$ activation	Mosley (1996); Rubin and McLeod (1994)
<5 s	↑ Protein kinase C activation	Jones et al. (1995)
6–20 min	↑ Enzyme activity G6PD, alkaline phosphatase	Skerry et al. (1989); Dodds et al. (1993)
5–15 min	↑ Prostaglandin expression	Rawlinson et al. (1991, 1993)
15 min	egr expression	Dolce et al. (1996)
<15 min	↑ NO production	Pitsillides et al. (1995)
48 h	Osteoblast proliferation, matrix synthesis	Pead et al. (1988)
>96 h	Mineralized bone formation	Many

G6PD, glucose-6-phosphate dehydrogenase; egr, ; NO, nitric oxide.

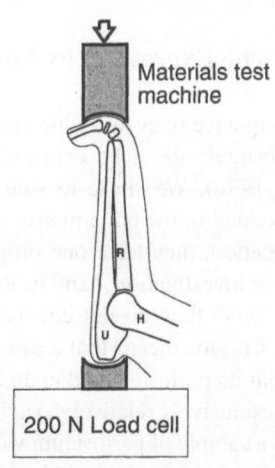

Fig. 3. The model for loading the rat ulna in vivo. *H*, humerus; *R*, radius; *U*, ulna. Axial compression causes the already curved bone to bend more, inducing strains which are maximal midway between the loading cups and distant from any point of pressure

characterize responses to different magnitudes and rates of loading and to study the reduced effectiveness of loading in ovariectomized animals compared with intact females (Mosley 1996). Because the sites of application of load are well away from the sample region, there appears to be insignificant complication with inflammatory responses. In old animals loading is capable of inducing bone formation on a quiescent surface (Torrance et al. 1994), but in younger rats, where one periosteal surface undergoes resorption as part of a modeling drift, the loading induces a sequence of changes beginning with inhibition of formation and culminating with formation, mimicking the events of the reversal phase of remodeling (Hillam and Skerry 1995). We can be confident that the loads we apply are close to the physiological range experienced by the animal, as we have measured strains in vivo in similarly sized animals and then performed load strain calibration studies in vitro. Interestingly, subsequent studies have also shown that, if instead of loading within the physiological range, the bone is fatigue-loaded on a single occasion, it is possible to induce intracortical haversian remodeling in the ulnar cortex, providing a suitable rodent model for that process (Bentonila et al. 1997).

8.6 Regulation of Gene Expression by Loading

Having established a sequence of events which follow loading of the rat ulna, we chose to investigate the gene expression of osteocytes after loading (Mason et al. 1996). We chose to study these cells because, although they are embedded in the bone matrix and are inaccessible to many forms of investigation, they have one property that makes them very appropriate for such investigation, namely, that they are fixed in the matrix and in primary bone they do not coexist with other cell types within that tissue (Fig. 4). This means that a sample of cortical bone, if isolated thoroughly from its periosteal and endosteal surfaces, contains cells which are almost exclusively relatively synchronized and quiescent osteocytes. In contrast, a sample of periosteum would contain numerous different cell types and many of those would be at different stages of the cell cycle. To isolate material from a small sample of cordial bone might seem difficult, but the method was simple. After thorough removal of the periosteum, the bone was snap frozen and mounted for cutting of

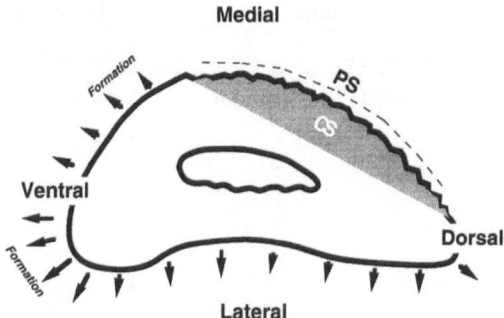

Fig. 4. Cross-section of the rat ulna midshaft. *CS*, cortical sample site used for PCR/DRD (differential RNA display) studies; *PS*, periosteal sample site. *Arrows* denote bone formation; *dashed line*, zone of bone resorption induced to form bone in response to loading

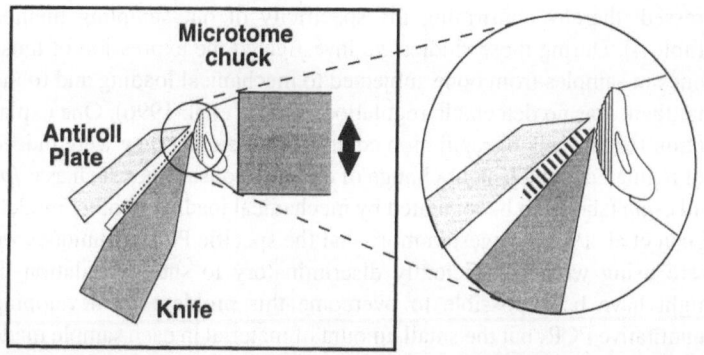

Fig. 5. Sampling method for PCR/DRD (differential RNA display) studies. Tangential cryostat sections accumulate under the antiroll plate, and as long as the removal of periosteum is thorough and the sectioning does not penetrate the marrow cavity, samples contain only cortical cells

tangential cryostat sections from the surface where loading induces stimulation of bone formation (Fig. 5). Although the samples involved were very small, they were amenable to analysis using PCR techniques, and we demonstrated the presence of several genes known to be expressed by osteocytes, and the absence of others which are not ex-

Table 4. Genes expressed in cortex and periosteum of the rat ulna

Gene	Osteocyte		Periosteum	
	Loaded	Control	Loaded	Control
β-actin	+	+	+	+
Osteocalcin	+	+	+	+
Connexin 43	+	+	+	+
Interleukin-1	−	−	−	−
TNF-α	−	−	+	+
TRAP	−	−	+	+
c-*fos*	+	+	+	+
c-*jun*	+	+	+	+

TNF-α, tumor necrosis factor-α; TRAP, tartrate-resistant acid phosphatase.
Using PCR, there is no apparent regulation by mechanical loading.

pressed, thereby confirming the specificity of our sampling method
(Table 4). During these studies we investigated the expression of those
genes in samples from bone subjected to mechanical loading and found
that there was no detectable regulation (Mason et al. 1996). One explanation for this lack of regulation could have been that they were indeed
not regulated, but a lack of change of early response genes such as c-*fos*
and c-*jun* (shown to be regulated by mechanical loading in other models
(Lean et al. 1995) suggested more that the specific PCR techniques we
were using were insufficiently discriminatory to show regulation. It
might have been possible to overcome this problem by developing
quantitative PCR, but the small amount of material in each sample made
such an approach difficult. Our strategy instead was to use differential
RNA display (DRD), which allows amplification of all sequences in a
sample, so that comparison of sequencing gels allows determination of
regulated genes which can then be excised and identified (Liang and
Pardee 1992) (Fig. 6). DRD techniques can be very elegant but are
prone to false positive results, suggesting regulation where none occurs.
Several criteria can assist with the minimization of false positives, and
these relate mainly to the specificity of the sample and the stimulus to be
investigated. From that viewpoint, the use of osteocytes for our studies
was ideal, as they were not able to cycle or perform many of the normal
actions of cells elsewhere because of their position inside the bone

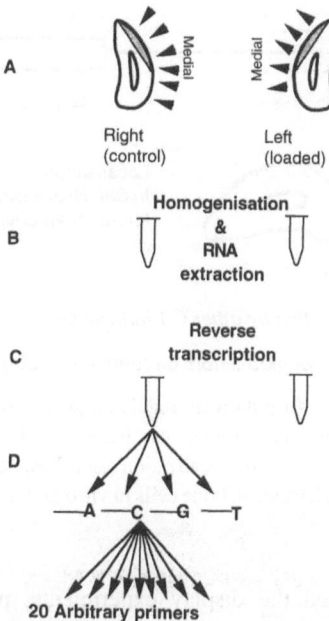

Fig. 6. Protocol for DRD (differential RNA display) studies. Samples of osteocytes from loaded and control bones are homogenized and RNA is extracted and reverse transcribed before amplification with sets of anchored primers to generate approximately 80 reactions per sample

matrix. Comparison of samples from osteocytes from left and right legs of the same animal eradicated differences in gene expression due to individual variation in the genotype of the animal and also any systemic fluctuations in hormones which might be different between individuals. The signal we applied to the cells in one bone was specific to that bone because we knew it did not affect bone formation in the other leg. Although we studied the effects of a period of loading a few hours after the application of the stimulus, we knew that the repeated application of the same stimulus resulted in inhibition of bone resorption and the stimulation of formation (Hillam and Skerry 1995). It is advisable to run duplicate samples in all display studies, when there is sufficient material, and only to consider as regulated those bands which are absent in one pair of lanes and present in the other pair.

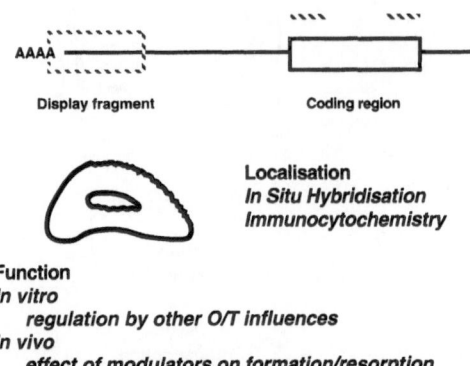

Fig. 7. Characterization of putatively regulated genes. Where sequences share homology with known genes, primers can be designed to the coding region of the molecule. Further studies on localization are followed by use of moderators of function to assess relevance in bone cells in vitro and in vivo. *O/T*, osteotropic

Having performed the display experiments, putatively regulated bands are excised from the sequencing gels and cloned and sequenced. Where sequences are compared with database information and found to be unknown, further study can be protracted as full length sequences have to be identified and probes and antibodies must be developed for further characterization and identification of function of regulated genes (Fig. 7). If sequences have homology with known genes, work can proceed much faster as it is possible to determine whether the gene in question is of interest in the context of a loading response. Many early response genes would be expected to be regulated by loading but it may not be productive to pursue such leads. What is more interesting is the situation in which a gene is identified that has some regulatory function in another tissue and that has not been identified in bone before. In that circumstance, there is a sequence of investigations which should be performed to characterize the response further.

If the gene has been cloned, it is possible to use PCR with specific primers directed to the coding or other regions of the molecule in order to characterize expression further. DRD generates products which are often in the untranslated region (UTR) of the gene, and with knowledge of the full sequence in another tissue more information can be obtained quickly by designing specific PCR primers. If coding region mRNA is expressed, then localization studies are valuable, in conjunction with Northern analysis where possible to confirm regulation of expression. In some circumstances, such as the use of the cryostat section DRD technique, there is insufficient material in even pooled samples for northern analysis. If a gene has been studied in other systems there are often antibodies available for use in western blotting and immunolocalization studies, and the availability of reagents makes the follow-up studies after DRD significantly more straightforward.

8.7 Mechanically Regulated Glutamate Transport in Bone

As an example of the process described above, we have performed several experiments to identify mechanically regulated genes in bone. The one which has been the most successful (Mason et al. 1997) showed that loading was associated with regulation of expression of the gene for a glutamate/aspartate transporter (GLAST) identified previously in the central nervous system (Storck et al. 1992). In the CNS, glutamate transporters are involved in reuptake of transmitters after synaptic transmission has occurred (Fig. 8). Synaptic transmission is the process in which apposed but unconnected cells communicate by paracrine release of neurotransmitters. Specialized glutamate vesicles are clustered as a readily releasable pool on the active zone of a presynaptic cell, with further reserve vesicles in the cytoplasm. In response to depolarization, glutamate is released into the synaptic cleft, where it diffuses across to the glutamate receptors on the postsynaptic cell (Hollmann and Heinemann 1994). Receptors are anchored in high density in the synaptic membrane by specific membrane clustering proteins (PSD-95 for NMDA receptors and GRIP for AMPA receptors (Gomperts 1996; Sheng and Kim 1996). In response to binding, the receptor channels open and there is rapid influx of calcium ions, which initiate secondary processes in the postsynaptic cell and affect synaptic plasticity by interactions with CAM kinase II and other enzymes associated with the carboxy terminus of the receptor and the clustering proteins. GLAST acst to buffer synaptic glutamate levels and is involved in reuptake into presynaptic cells and astrocytes.

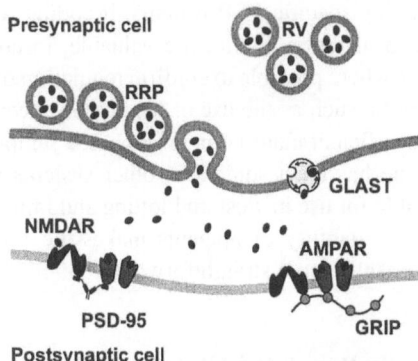

Fig. 8. Synaptic signaling in the CNS. Specialized glutamate vesicles are clustered as a readily releasable pool (*RRP*) on the active zone of a presynaptic cell, with further reserve vesicles (*RV*) in the cytoplasm. In response to depolarization, glutamate is released into the synaptic cleft, where it diffuses across to the glutamate receptors on the postsynaptic cell. Receptors are anchored in high density in the synaptic membrane by specific membrane clustering proteins (PSD-95 for NMDA receptors and GRIP for AMPA receptors). In response to binding, the receptor channels open and there is rapid influx of calcium ions, which initiate secondary processes in the postsynaptic cell and affect synaptic plasticity by interactions with CAM kinase II and other enzymes associated with the carboxy terminus of the receptor and the clustering proteins. Glutamate transporters (*GLAST*) act to buffer synaptic glutamate levels and are involved in reuptake into presynaptic cells and astrocytes

When we identified GLAST expression in bone it became clear that a similar mode of intercellular communication between bone cells may occur as does at synapses in the CNS. Our studies have therefore focused on investigation of expression of glutamate receptors and other parts of the system described above as well as functional assays to determine the role of such signaling in bone cells. To date the evidence for such a role has become steadily more compelling (Skerry et al. 1996) and it appears that both bone formation and resorption in vitro can be regulated by modulation of glutamate receptor actions in bone cells.

8.8 Conclusion

In this chapter I have attempted to illustrated the general principles and some of the specific issues involved in the use of mechanical loading experiments to uncover new targets for stimulation of bone formation as a treatment for osteoporosis. Naturally such a topic is very large and cannot be dealt with fully in the space available. However, the identification of a novel means of influencing bone cell activity by use of drugs to modulate bone cell glutamate signaling suggests that our approach has merit.

The dramatic growth of sophisticated techniques in molecular biology has led to prodigious advances in biology generally, and in bone physiology specifically. The results of subtractive library techniques, high throughput screening and single cell PCR (for example) are only just beginning to become widely known. These data are the result of work which was performed some years ago, so it is inconceivable that in the near future there will not be still more remarkable advances in understanding, leading to the development of clinical modalities to treat bone loss.

Acknowledgements. Our studies have been funded by Arthritis and Rheumatism Council, Action Research, the BBSRC, the Medical Research Council, the Nuffield Foundation, Smith and Nephew, and the Wellcome Trust, and I would like to thank them for that support. Many colleagues have been involved in these experiments and I would like to acknowledge particularly Professor Lance Lanyon and Dr John Mosley in London, Drs Bev Fermor, Richard Hillam and Debbie Mason in Bristol, Professor Larry Suva and Sharon Stueckle in Boston and all the members of my group in York without whom none of this work would have occurred.

References

Bassey EJ, Ramsdale SJ (1994) Increase in femoral bone density in young women following high-impact exercise. Osteoporosis Int 4:72–75
Bassey EJ, Ramsdale SJ (1995) Weight-bearing exercise and ground reaction forces – a 12-month randomized controlled trial of effects on bone-mineral density in healthy postmenopausal women. Bone 16:469–476

Bentonila V, Hillam RA, Skerry TM, Boyd TM, Fyhrie D, Schaffler MB (1997) Activation of intracortical remodeling in adult rat long bones by fatigue loading. Trans Orthop Res Soc 22:578

Dodds RA, Ali NN, Pead MJ, Lanyon LE (1993) Early loading-related changes in the activity of glucose 6-phosphate dehydrogenase and alkaline phosphatase in osteocytes and periosteal osteoblasts in rat fibulae in vivo. J Bone Miner Res 8:261–267

Dolce C, Kinniburgh AJ, Dziak R (1996) Immediate early-gene induction in rat osteoblastic cells after mechanical deformation Arch Oral Biol 41:1101–1108

Doty SB (1981) Morphological evidence of gap junctions between bone cells. Calcif Tiss Int 33:509–512

Duncan RL, Turner CH (1995) Mechanotransduction and the functional-response of bone to mechanical strain. Calcif Tiss Int 57:344–358

Gomperts SN (1996) Clustering membrane proteins: Its all coming together with the PSD-95/SAP90 protein family. Cell 84:659–662

Hert J, Skelenska A,Liskova M (1971) Continuous and intermittent loading of the tibia in rabbit. Folia Morphol 19:378–387

Hillam RA, Mosley JM, Skerry TM (1994) Regional differences in bone strain. Bone Miner 25:S1–32 (Abstract)

Hillam RA, Skerry TM (1995) Inhibition of bone resorption and stimulation of formation by mechanical loading of the modeling rat ulna in vivo. J Bone Miner Res 10:683–689

Hollmann M, Heinemann S (1994) Cloned glutamate receptors. Ann Rev Neurosci 17:31–108

Hylander WL, Johnson KR, Crompton AW (1987) Loading patterns and jaw movements during mastication in macaca-fascicularis – a bone-strain, electromyographic, and cineradiographic analysis. Am J Phys Anthropol 72:287–314

Hylander WL, Johnson KR (1997) In vivo bone strain patterns in the zygomatic arch of macaques and the significance of these patterns for functional interpretations of craniofacial form. Am J Phys Anthropol 102:203–232

Jones DB, Leivseth G,Tenbosch J (1995) Mechano-reception in osteoblast-like cells. Biochem Cell Biol 73:525–534

Jones SJ, Gray C, Sakamaki H, Arora M, Boyde A, Gourdie R, Green C (1993) The incidence and size of gap-junctions between the bone-cells in rat calvaria. Anat Embryol 187:343–352

Keila S, Pitaru S, Grosskopf A, Weinreb M (1994) Bone-marrow from mechanically unloaded rat bones expresses reduced osteogenic capacity in-vitro J Bone Miner Res 9:321–327

Lean JM, Jagger CJ, Chambers TJ, Chow JWM (1995) Increased insulin-like growth factor I mRNA expression in rat osteocytes in response to mechanical stimulation. Endocrinol Metab 268:318–327

Liang P, Pardee AB (1992 Differential display of eukaryotic messenger-RNA by means of the polymerase chain-reaction. Science 257:967–971

Mason DJ, Hillam RA, Skerry TM (1996) Constitutive in vivo mRNA expression by osteocytes of β-actin, osteocalcin, connexin-43, IGF I, c-fos and c-jun, but not TNFa or tartrate resistant acid phosphatase. J Bone Miner Res 11:3 350–357

Mason DJ, Suva LJ, Genever PG, Patton AJ, Stueckle S, Hillam RA, Skerry TM (1997) Mechanically regulated expression of a neural glutamate transporter in bone. A role for excitatory amino acids as osteotropic agents? Bone 20:199–205

Mosley JM (1996) The influence of mechanical load and oestrogen on the development of long bone architecture. PhD thesis, University of London

Mullender MG, Huiskes R (1997) Osteocytes and bone lining cells: which are the best candidates for mechano-sensors in cancellous bone? Bone 20:527–532

O'Connor JA, Lanyon LE, McFie HF (1982) The influence of strain rate on adaptive bone remodeling. J Biomech 15:767–781

Pead MJ, Skerry TM, Lanyon LE (1988) Direct transformation from quiescence to bone formation in the adult periosteum following a single brief period of bone loading. J Bone Miner Res 3:647–656

Pitsillides AA, Rawlinson SCF, Suswillo RFL, Bourrin S, Zaman G. Lanyon LE (1995) Mechanical strain-induced no production by bone-cells – a possible role in adaptive bone (re)modeling. FASEB J 9:1614–1622

Rawlinson SCF, El Haj AJ, Minter SL, Tavares IA, Bennett A, Lanyon LE (1991) Loading-related increases in prostaglandin production in cores of adult canine cancellous bone in vitro: a role for prostacyclin in adaptive bone remodeling? J Bone Miner Res 6:1345–1351

Rawlinson SCF, Mohan S, Baylink DJ, Lanyon LE (1993 Exogenous prostacyclin, but not prostaglandin E2, produces similar responses in both G6PD activity and RNA production as mechanical loading, and increases IGF-II release, in adult cancellous bone in culture. Calcif Tiss Int 53:324–329

Rubin CT, Lanyon LE (1987) Osteoregulatory nature of mechanical stimuli: function as a determinant for adaptive remodeling in bone. J .Orthop .Res 5:300–310

Rubin CT, McLeod KJ (1994) Promotion of bony ingrowth by frequency-specific, low-amplitude mechanical strain. Clin Orthop Rel Res 298:165–174

Rubin CT, Lanyon LE (1984) Dynamic strain similarity in vertebrates; an alternative to allometric limb bone scaling. J Theor Biol 107:321–327

Sheng M, Kim E (1996) Ion-channel associated proteins. Curr Opin Neurobiol 6:602–608

Skerry TM, Bitensky L, Chayen J, Lanyon LE (1989) Early strain-related changes in enzyme activity in osteocytes following bone loading in vivo. J Bone Miner Res 4:783–788

Skerry TM, Genever PG, Patton AJ, Grabowski PS, Stueckle S, Suva LJ (1996) Glutamate receptors in bone-cells suggest a paracrine role for excitatory amino-acids in regulation of the skeleton. J Bone Miner Res 11:202

Skerry TM (1997) Mechanical loading and bone: What sort of exercise is beneficial to the skeleton? Bone 20:179–181

Skerry TM, Peet NM (1997) "'Unloading" exercise increases bone formation in rats. J Bone Miner Res 12:6

Smith EL, Gilligan C (1991) Physical-activity effects on bone metabolism. Calcif Tiss Int 49:S50–S54

Smith EL, Raab DM (1986) Osteoporosis and physical activity. Acta Med Scand Suppl 711:149–156

Storck T, Schulte S, Hoffman K, Stoffel W (1992) Structure, expression and functional analysis of a Na+ dependent glutamate/aspartate transporter from rat brain. Proc Natl Acad Sci USA. 89:10955–10959

Thomas GP, ElHaj AJ (1996) Bone-marrow stromal cells are load responsive in-vitro. Calcif Tiss Int 58:101–108

Torrance AG, Mosley JM, Suswillo RFL, Lanyon LE (1994) Noninvasive loading of the rat ulna in vivo induces a strain related modeling response uncomplicated by trauma of periosteal pressure. Calcif Tiss Int 54:241–247

Turner CH, Owan I, Takano Y (1995a) Mechanotransduction in bone – role of strain-rate Endocrinol Metab 32:E438–E 442

Turner CH, Takano Y, Owan I (1995b) Aging changes mechanical loading thresholds for bone-formation in rats. J Bone Miner Res 10:1544–1549

Vaes G (1988) Cellular biology and biochemical mechanism of bone resorption. A review of recent developments on the formation, activation, and mode of action of osteoclasts. Clin Orthop 231:239–271

Wozasek GE, Simon P, Redl H, Schlag G (1994) Intramedullary pressure changes and fat intravasation during intramedullary nailing – an experimental-study in sheep. J Trauma 36:202–207

9 Amplification of the Osteogenic Stimulus of Load-Bearing as a Logical Therapy for the Treatment and Prevention of Osteoporosis

L.E. Lanyon

9.1 Osteoporosis as a Failure of Strain-Related Control of Bone Architecture

Normal bone architecture implies an ability to withstand functional load-bearing without damage. It indicates a skeleton with normal adaptability to mechanical loading and a normal loading history. There is a substantial literature documenting instances in which changes in loading history are reflected by changes in bone structure. There are also instances in which the normal structure:functional relationship is achieved but not maintained despite apparently normal loading. The most significant single instance of this is post-menopausal osteoporosis.

Bone cells are presumed to ensure the structural competence of the skeleton by adjusting bone architecture so that functional loading engenders "target" levels of strain in the bone tissue (Lanyon 1987, 1992). Failure to achieve these target strains is accompanied by bone loss, whereas if they are exceeded adaptive hypertrophy normally ensues. To regulate bone architecture in this strain-related homeostatic manner requires bone cells to be able: (1) to measure the strains they experience, (2) to compare the strains experienced with the target levels determined (genetically?) to be appropriate for that location, and (3) to stimulate and control any adaptive response necessary to adjust the mass and architecture of the bone to eliminate any discrepancy between the two. Osteoporosis, with its structurally inappropriate loss of bone despite continued functional loading, is by definition a failure of this mechanism.

Although the etiology of osteoporosis, and the means to treat it, should logically be sought in the mechanisms involved in bone cells responses to mechanical strain, the established treatments for osteoporosis are empirical. Hormone replacement therapy replaces estrogen, the withdrawal of which is evidently associated with rapid bone loss. Bisphosphonates and calcitonin directly reduce resorption without which bone loss cannot occur.

Recently, attention has focused on the possibility of treating and preventing osteoporosis through physical exercise. This approach seeks to harness the mechanisms of strain-related (re)modeling without in any way understanding what they are.

9.2 In Vivo Evidence of Loading-Related Control of Bone Architecture

What knowledge we have of strain-related remodeling has been primarily derived from loading experiments in whole animals in vivo. Currently there are only four in vivo loading models contributing to the literature. These are: (1) the functional rat ulna model, which is valuable but not ideal (Torrance et al. 1994; Mosley et al. 1997); (2) the isolated avian ulna and radius models, which are cumbersome, invasive and nonmammalian (Rubin and Lanyon 1984, 1985); (3) the rat tibial four-point bending model, which we consider flawed since load is applied through the periosteum, which in consequence elaborates woven bone at the points of pressure (Turner et al. 1991); and (4) the rat tail model, which is invasive, uses very high strains and relates primarily to cancellous bone (Chambers et al. 1993).

Despite their disadvantages, all these models, and the various osteotomy and artificial loading models which preceded them (i.e., Goodship et al. 1979; Lanyon et al. 1982; OConnor et al. 1982), have contributed to our understanding of the osteoregulatory components of the strain wave form (such as strain magnitude, frequency, and strain rate). They show that bone (re)modeling may be profoundly influenced by short periods of loading at daily intervals and that the osteogenic potential of the response is related to the peak strain magnitude, the maximum strain rate, the component frequencies of the strain wave form and the mismatch between the bones' habitual strain distributions and those to which they are currently exposed.

The finding, in the avian ulna preparation, of the short period of time necessary to produce a maximal adaptive response (Rubin and Lanyon 1984) transformed our view of the nature of the mechanical stimulus responsible for ensuring structurally appropriate bone architecture. Up until that study it had been generally assumed that the predominant stimulus was produced by the predominant activity. Thus if an individual ran for 3 h a day their skeletal architecture would reflect the stimulus of the loads engendered by the process of running. If that were the case then bone cells would not need to respond as swiftly as they do. That they should respond so swiftly gave rise to the idea that bone cells were less responsive to the loads of repetitive normal activities than those generated during transient loading accidents. This implies that adaptive control of bone architecture is driven by the strains encountered during loading errors and not those generated during controlled activities, which in any case are usually more sedate (Lanyon 1992).

By using a model in which the sheep calcaneous was either protected from or exposed to functional load-bearing, we have shown that the strains of controlled walking are in fact as ineffective in preventing bone loss as disuse (Skerry and Lanyon 1995). This supports the hypothesis that control of bone mass is preferentially influenced by unusual and diverse strain distributions and error type wave forms.

9.3 Exercise and Bone Mass

The suggestion that bone mass is preferentially influenced by the strains
encountered during unusual loading events is supported by a number of
exercise studies which show that athletes involved in ballistic sports like
squash and gymnastics have a higher bone mineral density (BMD) than
physically active referents, whereas the BMD of those involved in
cycling and swimming is no different (Heinonen et al. 1995). Animal
and human studies also show that the skeleton is most receptive to
mechanical stimulation during growth (Kannus et al. 1995), but nevertheless high impact exercise can produce demonstrable increases in
bone mass in women aged 35–45 (Heinonen et al. 1996).

The implication of these studies is clear. The type of exercise necessary to stimulate increases in bone mass in people suffering from, or at
risk from, osteoporosis is just that which is too dangerous to be employed as a therapy for these people. However, the strain-related influence on bone remodeling is the only one capable of stimulating a
specific structurally appropriate response. Since increased loading cannot be employed, the therapeutic strategy of choice must be to amplify
the osteogenic/antiresorptive effects of safe levels of mechanical loading. To do this requires knowledge of the mechanisms involved in bones'
normal responses to such loading.

9.4 The Mechanisms of Strain-Related Adaptive (Re)Modeling

Bone cells of two lineages are involved in bone modeling and remodeling: one gives rise to osteoblasts, osteocytes and lining cells and the
other to osteoclasts. Of the adult cells of these lineages, the osteocytes,
lining cells and osteoblasts are the permanent residents and are in the
best position to respond to strains in the bone tissue (Lanyon 1993). Of
these, the greater number and the best placed are the osteocytes. However, it is evident that both osteoblasts and osteocytes respond to strain
in their immediate locality (Skerry et al. 1989; Cheng et al. 1994), and
it is likely that they also control osteoclast activity and perhaps osteoclast recruitment.

Osteocytes in situ in superfused cores of canine cancellous bone respond to loading by the production of prostacyclin (PGI2). Osteoblasts produce PGI_2 and PGE_2 (Rawlinson et al. 1991). Both osteocytes and osteoblasts produce nitric oxide(NO) (Pitsillides et al. 1995a), but osteocytes produce more NO per cell than osteoblasts (Pitsillides et al. 1995b). Both NO and PG production appear to be obligatory for a subsequent osteogenic response (Fox et al. 1996; Turner et al. 1996; Pead and Lanyon 1989; Chow and Chambers 1994). The predominant NO synthase responsible for NO production is the constitutive endothelial isoform eNOS (Hukkanen et al. 1997; Chow et al. 1996). Interestingly, calvarial cells do not show the same response to loading either in situ or in monolayer culture as cells in long bones (Rawlinson et al. 1995). We do not therefore consider it safe to extrapolate between the responses to mechanical loading of cells derived from calvaria and those derived from long bones.

In externally loaded explants of rat ulnae osteocytes and osteoblasts show a uniform sensitivity to exogenous PGE_2 and PGI_2 in terms of increased glucose 6-phosphate dehydrogenase (G6PD) activity. However, both in vivo (Skerry et al. 1989) and in vitro (Cheng et al. 1994), their (indomethacin-blockable) increase in G6PD activity in response to loading is directly related to their local strain magnitude. This suggests that bone cells throughout the structure are uniformly sensitive to strain, to which they respond via strain magnitude-related production of prostanoids and NO.

These strain-related responses in osteoblasts and osteocytes in rat ulna explants can be blocked by selective ion channel blockers (Rawlinson et al. 1996). Gadolinium chloride, which blocks stretch-activated cation channels, eliminates the strain-related increase in G6PD activity in osteocytes but only reduces it in osteoblasts. Nifedipine, which blocks voltage-dependent calcium channels, inhibits strain-related increase in G6PD activity in osteoblasts but has no effect in osteocytes.

In explants loading causes rapid transient increase in *c-fos* expression followed closely by increased expression of insulin-like growth factor (IGF)-II (Zaman et al. 1994). In superfused cancellous bone cores exogenous PGI_2 (normally produced by loading) increases IGF-II in the medium but not IGF-I (Rawlinson et al. 1993). This implication of IGF-II in our experiments contrasts with in situ hybridization findings from others that loading up-regulates osteocyte IGF-I (Lean et al. 1995).

If osteoblasts were the only resident cell type involved in the adaptive responses to strain it would be easy enough to consider that they produce prostanoids, NO and growth factors in a local strain magnitude-related manner and that these act in an autocrine fashion to stimulate local strain magnitude-related new bone formation. This explanation avoids a number of relevant questions including: (1) why do osteocytes respond to strain with NO and PG production as vigorously as osteoblasts if they are not usefully involved in the strain-related response, and (2) how would a local autocrine response such as this control appropriate adaptive (re)modeling when the adaptive response throughout the bone structure does not occur in a local strain magnitude-related manner?

Our studies using the noninvasive method of loading the rat ulna in vivo through the olecranon and flexed carpus (Torrance et al. 1994) have confirmed, what should have been obvious from thought alone, that whole bones respond to loading as structures, modifying their longitudinal curvature, area and cross-sectional shape by coordinated modulation of formation and resorption at different locations (Mosley et al. 1997). This means that the adaptive (re)modeling response to loads cannot be directly derived from the initial strain-related response of osteocytes and osteoblasts which measure strain, but from their responses after subsequent modification and /or processing.

The osteocyte:osteoblast network is well arranged for such processing since each osteocyte is connected via gap junctions to its neighbors and to osteoblasts on the bone surfaces. Whether this network functions in this way is not known. Only ions and other small molecules can pass through gap junctions, the most likely being Ca^{2+} and inositol trisphosphate (IP_3). The IGFs which are produced by osteocytes in response to loading are far too large to traverse the gap junctions with or without their binding proteins. Assuming that these growth factors are involved in the adaptive process, they are presumably transported from the osteocytes to the surfaces on which modeling or remodeling occurs in the extracellular fluid. Although the extracellular fluid is to some extent pumped by functional strains in a strain gradient-related manner, this process seems rather casual to produce such a precise result. An hypothesis that we find preferable is that processing of the raw strain data occurs in the osteocyte:osteoblast network by means which we do not yet understand but possibly involving neurotransmitters such as glutamate, whose presence has been detected but whose role and mechanism of action remain to be determined (Mason et al. 1997).

Osteocytes can stay alive in situ for over 70 years (Marotti et al. 1987), yet a small proportion of them are apoptotic during growth. The number of apoptotic osteocytes is increased in the absence of estrogen (Tomkinson et al. 1997). Preliminary data in rats show that strain also reduces the number of apoptotic osteocytes, in this case exponentially with peak strain magnitude (Noble et al. 1997a, b). It is tempting to speculate that the products associated with apoptosis contribute to osteocyte signaling. What they signal is likely to be either the raw data of their response to the strains that they have just encountered, or strain data that has been processed both in relation to strains in other parts of the bone, and to systemic influences such as estrogen levels.

Elucidating the mechanism by which strain is converted into a relevant stimulus for (re)modeling is one of the great challenges in this field. The other is determining why withdrawal of estrogen renders the adaptive process so ineffectual that post-menopausal osteoporosis results.

9.5 The Role of Estrogen in Strain-Related (Re)Modeling

It is quite possible that the increasing incidence of osteoporosis may be contributed to by both the low general level of load-bearing of modern life and by a lack of specific (error-rich) components of loading with a high osteogenic potential. Although inadequate mechanical stimulation may contribute to the incidence and severity of osteoporosis, the rapid loss of bone at the menopause, and its stabilization by hormone replacement therapy (HRT), clearly implicates estrogen in its etiology. However, post-menopausal bone loss cannot simply be explained by the withdrawal of a direct effect of estrogen on bone cells, since, if this were the case, the absence of estrogen's contribution should be compensated for within the strain-regulated homeostatic mechanism by an increased stimulus due to the high strains associated with low bone mass (Cheng et al. 1996). For estrogen to have an effect that is not compensated for in this way implies that estrogen is involved in the mechanically adaptive response, which is thus, at least partially, disabled by its withdrawal.

The cascade of cellular events between loading and change in bone architecture involves a large number of stages, a great many of which are potentially affected by estrogen. Using explants of ulnae from rats we have shown that in females, but not males, there is synergism between the effects of estrogen and loading on both osteoblast proliferation and matrix production (Cheng et al. 1996). This synergism is not reproduced in primary cell culture where the effects of strain and estrogen are additive. However, in these cell cultures the estrogen antagonists tamoxifen and ICI 182,780 have been shown to reduce not only the cells' proliferative responses to estrogen but also those to strain (Damien et al. 1996, 1997). This implies involvement of the estrogen receptor in the early responses to strain. Such involvement would provide a rational explanation for the etiology of post-menopausal osteoporosis in which adaptability to mechanical loading is plainly deficient in the absence of estrogen when the estrogen receptor is likely to be down-regulated.

9.6 A Strategy for the Treatment and Prevention of Osteoporosis

1. Since load-bearing is the only functional influence on bone (re)modeling capable of stimulating a specific structurally appropriate response, the therapeutic strategy of choice should be to harness this influence.
2. The load-bearing employed to stimulate adaptive maintenance or increase in bone mass must not increase the danger of fracture. Since the most osteogenic loading regimens are those which are most dangerous they cannot be employed.
3. A successful strategy for the prevention and treatment of osteoporosis could be to amplify the osteogenic/ antiresorptive response to a safe level of loading. Many of the stages in bone cells responses to loading involve factors such as prostanoids, NO and growth factors that are not specific to bone and thus are not likely to be able to produce positive effects in bone without affecting other systems.
4. The estrogen receptor modulators tamoxifen and ICI 182,780 have been shown to be capable of blocking bone cells adaptive responses to strain. Other estrogen receptor modulators may be able to enhance this response. Selective estrogen receptor modulators (SERMs) have tissue specificity. Exploring the possibility of a SERM which can amplify the initial strain-related stimulus in bone

tissue without having adverse effects in others should be a high priority.

Acknowledgements. The work described here is currently supported by the Wellcome Trust, the Medical Research Council and the BBSRC.

References

Chambers TJ, Evans M, Gardner TN, Turner Smith A, Chow JW (1993) Induction of bone formation in rat tail vertebrae by mechanical loading. Bone Miner 20:167–178

Cheng MZ, Zaman G, Lanyon LE (1994) Estrogen enhances the stimulation of bone collagen synthesis by loading and exogenous prostacyclin, but not prostaglandin E_2, in organ cultures of rat ulnae. J Bone Miner Res 9:805–816

Cheng MZ, Zaman G, Rawlinson SCF, Suswillo RFL, Lanyon LE (1996) Mechanical loading and sex hormone interactions in organ cultures of rat ulna. J Bone Miner Res 11:502–511

Chow JWM, Chambers TJ (1994) Indomethacin has distinct early and late actions on bone-formation induced by mechanical stimulation. Am J Physiol 267:E 287–E292

Chow JWM, Chambers TJ, Fox SW (1996) Immunolocalization of nitric oxide synthase in bone cells. J Bone Miner Res 11:M354

Damien E, Price JS, Suswillo RFL, Lanyon LE (1996) Estrogens role in regulating the response of bone-cells to mechanical strain. J Bone Miner Res 11:9

Damien E, Price JS, Lanyon LE (1997) Involvement of the estrogen receptor in bone cells' adaptive response to mechanical strain. J Bone Miner Res 12:F230

Fox SW, Chambers TJ, Chow JW (1996) Nitric oxide is an early mediator of the increase in bone formation by mechanical stimulation. Am J Physiol 270:E955–960

Goodship AE, Lanyon LE and McFie H (1979 Functional adaptation of bone to increased stress. J Bone Joint Surg 61-A :539–546

Heinonen A, Oja P, Kannus P, Sievanen H, Haapasalo H, Manttari A, Vuori I (1995) Bone mineral density in female athletes representing sports with different loading characteristics of the skeleton. Bone 17:197–203

Heinonen A, Kannus P, Sievanen H, Oja P, Pasanen M, Rinne M, Uusi Rasi K, Vuori I (1996) Randomised controlled trial of effect of high-impact exercise on selected risk factors for osteoporotic fractures Lancet 348:1343–1347

Hukkanen M, Pitsillides AA, Rawlinson SCF, Suswillo RFL, Zaman G, Platts LAM, Polak JM, Lanyon LE (1997) A putative role for osteocyte e-NOS-derived NO in adaptive bone remodeling. J Path 181:A42

Kannus P, Haapasalo H, Sankelo M, Sievanen H, Pasanen M, Heinonen A, Oja P, Vuori I (1995) Effect of starting age of physical activity on bone mass in the dominant arm of tennis and squash players. Ann Intern Med 123:27–31

Lanyon LE Goodship AE Pye CJ and McFie H (1982) Mechanically adaptive bone remodeling. A quantitative study on functional adaptation in the radius following ulna osteotomy in sheep. J. Biomechanics 15:141–154

Lanyon LE (1987 Functional strain in bone tissue as an objective and controlling stimulus for adaptive bone remodeling. J Biomechanics 20:1083–1093

Lanyon LE (1992) The success and failure of the adaptive response to functional load-bearing in averting bone fracture. Bone 13:S17–S21

Lanyon LE (1993) Osteocytes, strain detection, bone modeling and remodeling. Calcif Tissue Int 53:S102–S107

Lean JM, Jagger CJ, Chambers TJ, Chow JW (1995) Increased insulin-like growth factor I mRNA expression in rat osteocytes in response to mechanical stimulation. Am J Physiol 268:E318–E327

Marotti G Balli R Remaggi F Farneti D (1987) Morphofunctional study on osteocytes in normal auditory ossicles. Acta Otorhinol Ital 7:347–363

Mason DJ, Suva LJ, Genever PG, Patton AJ, Steuckle S, Hillam RA, Skerry TM (1997) Mechanically regulated expression of a neural glutamate transporter in bone: A role for excitatory amino acids as osteotropic agents? Bone 20:199–205

Mosley JR, March BM, Lynch J, Lanyon LE (1997) Strain magnitude related changes in whole bone architecture in growing rats. Bone 20:191–198

Noble BS, Stevens H, Mosley JR, Pitsillides AA, Reeve J, Lanyon LE (1997a) Bone loading changes the number and distribution of apopotic osteocytes in cortical bone. J Bone Miner Res 12:S111

Noble BS, Stevens H, Mosley JR, Pitsillides AA, Reeve J, Lanyon LE (1997b) Osteocyte apoptosis and functional strain in bone. J Bone Miner Res 12:1520

O'Connor JA, Lanyon LE, McFie H (1982) The influence of strain rate on adaptive bone remodeling. J Biomechanics 15:767–781

Pead MJ, Lanyon LE (1989) Indomethacin modulation of load-related stimulation of new bone formation in vivo. Calcif Tissue Int 45:34–40

Pitsillides AA, Rawlinson SC, Suswillo RF, Bourrin S, Zaman G, Lanyon LE (1995a) Mechanical strain-induced NO production by bone cells: a possible role in adaptive bone (re)modeling? FASEB J 9:1614–622

Pitsillides AA, Rawlinson SCF, Suswillo RFL, Zaman G, Nijwiede PJ, Lanyon LE (1995b) Mechanical strain-induced NO production by osteoblasts and osteocytes. J Bone Miner Res 10:S217

Rawlinson SC, el Haj AJ, Minter SL, Tavares IA, Bennett A, Lanyon LE (1991) Loading-related increases in prostaglandin production in cores of adult canine cancellous bone in vitro: a role for prostacyclin in adaptive bone remodeling? J Bone Miner Res 6:1345–1351

Rawlinson SC, Mohan S, Baylink DJ, Lanyon LE (1993) Exogenous prostacyclin, but not prostaglandin E_2, produces similar responses in both G6PD activity and RNA production as mechanical loading, and increases IGF-II release, in adult cancellous bone in culture. Calcif Tissue Int 53:324–329

Rawlinson SC, Mosley JR, Suswillo RF, Pitsillides AA, Lanyon LE (1995) Calvarial and limb bone cells in organ and monolayer culture do not show the same early responses to dynamic mechanical strain. J Bone Miner Res 10:1225–1232

Rawlinson SC, Pitsillides AA, Lanyon LE (1996) Involvement of different ion channels in osteoblasts' and osteocytes' early responses to mechanical strain. Bone 19:609–614

Rubin CT, Lanyon LE (1984) Regulation of bone formation by applied dynamic loads. J Bone Joint Surg Am 66:397–402

Rubin CT, Lanyon LE (1985) Regulation of bone mass by mechanical strain magnitude. Calcif Tissue Int 37:411–417

Skerry TM Bitensky L Chayen J Lanyon LE (1989) Early strain-related changes in enzyme activity in osteocytes following bone loading in vivo. J Bone Miner Res 4:783–788

Skerry TM, Lanyon LE (1995) Interruption of disuse by short duration walking exercise does not prevent bone loss in the sheep calcaneus. Bone 16:269–274

Tomkinson A, Reeve J, Shaw RW, Noble BS (1997) The death of osteocytes by apoptosis in human bone is observed following estrogen withdrawl by GnRH analogs. J Clin Endo and Metab 82:3128–3135

Torrance AG, Mosley JR, Suswillo RF, Lanyon LE (1994 Noninvasive loading of the rat ulna in vivo induces a strain-related modeling response uncomplicated by trauma or periostal pressure. Calcif Tissue Int 54:241–247

Turner CH, Akhter MP, Raab DM, Kimmel DB, Recker RR (1991 A noninvasive, in vivo model for studying strain adaptive bone modeling. Bone 12:73–79

Turner CH, Takano Y, Owan I, Murrell GA (1996) Nitric oxide inhibitor L-NAME suppresses mechanically induced bone formation in rats. Am J Physiol 270:E634–E639

Zaman G, Suswillo RFL, Cheng MZ, Lanyon LE (1994) Effect of strain and strain-related prostanoids on mRNA expression of c-fos, IGF-I, IGF-II and $TGF\beta_1$. J Bone Miner Res 9:S303

10 Novel Aspects of Parathyroid Hormone/Parathyroid Hormone-Related Protein Hormone-Receptor Interactions

L.J. Suva

10.1 Introduction

The primary objective of the investigation of the fundamental nature of ligand-receptor interactions is elucidation of the mechanism of ligand recognition and receptor activation. In the case of parathyroid hormone

(PTH) and PTH-related protein (PTHrP), which have been well studied from the ligand side of this interaction for many years, the partnership of chemistry and pharmacology have provided important new insight into the mode of interaction of the ligand with its receptor and has guided the design of new agonists and antagonists as well as analogues with specialized applications (Chorev and Rosenblatt 1996; Gardella et al. 1994, 1996). However, it is also safe to say that despite such intense study, the considerable achievements have not included the "holy grail" of the identification of agonists with biological activity superior to those of parent peptide, namely PTH-(1–34) (Chorev and Rosenblatt 1996). In a parallel effort, modern biology and molecular biology techniques have resulted in the cloning of the receptors for PTH (Schipani et al. 1993; Adams et al. 1995; Usdin et al. 1995) and the development of sophisticated biological assay systems (Pines et al. 1994, 1996; Adams et al. 1995) for PTH action. Advances in chemical synthesis (Chorev and Rosenblatt 1994; Gardella et al. 1994, 1996) and the high-resolution solution NMR structures of the ligands (Pellegrini et al. 1997) have enabled the production of specialized, biologically active analogues with which to probe the bimolecular interaction between ligand and receptor (Nakamoto et al. 1995). Such investigations should enable us to gain structural insights into the requirements of both ligand and receptor for functional interaction and guide the development of rationally designed, small bioactive ligands.

From a purely pharmaceutical perspective, it is likely that the next major advancement in osteoporosis research will involve the development of a potent anabolic agent. PTH is one of the few bone-active agents proven to be anabolic in humans (Dempster et al. 1993) and so has the important criterion of validation already provided. In addition, the development of an anabolic agent which is mechanism-based is very attractive. PTH mediates its effects via a seven transmembrane (TM) receptor (PTH1 receptor) localized in bone and kidney (Dempster et al. 1993), which is coupled to both cAMP and intracellular calcium pathways (Abou-Samra et al. 1989; Iida-Klein et al. 1989; Partridge et al. 1981; Pines et al. 1994, 1996; Adams et al. 1995). Although it is not clear which of these pathways is responsible for the anabolic activity of intermittent PTH administration, the effects are certainly PTH1 receptor-mediated (Dempster et al. 1993; Pines et al. 1994, 1996).

We have recently developed an approach to study the interaction between PTH and the human PTH1 receptor, which utilizes direct contact domain identification (Adams et al. 1995; Nakamoto et al. 1995; Zhou et al. 1997; Bisello et al. 1998), together with molecular dynamics and molecular modeling. The combined power of these experimental approaches has enabled development of a molecular simulation of the ligand-receptor interface, which provides the first opportunity to rationally probe the hormone-receptor bimolecular interaction. These studies may allow the development of a new generation of PTH-based ligands with enhanced biological activity.

10.2 Methods

10.2.1 Peptide Synthesis

All peptides were synthesized by conventional solid phase methodology using Boc/HOBt/NMP chemistry on an Applied Biosystems 430 A peptide synthesizer. After hydrogen fluoride cleavage, peptides were purified by RP-HPLC (Nakamoto et al. 1995). Purity and structure of all synthesized peptides were confirmed by analytical RP-HPLC, amino acid analysis and electron spray mass spectrometry. The two ligands used in these studies K13 ([Nle^{8,18}, Lys^{13}(ε-pBz$_2$), Arg^{13,26,27},L-2-Nal^{23}, Tyr^{34}]bPTH-(1–34)NH$_2$), and Bpa-1 ([Bpa-1, Nle^{8,18}, Arg^{13,26,27}, L-2-Nal^{23}, Tyr^{34}]bPTH-(1–34)NH$_2$), are heavily substituted to make them resistant to the chemical and enzymatic cleavage protocols we employed. Radioiodination of [Nle^{8,18}, Tyr^{34}]bPTH-(1–34)NH$_2$ (or other ligands) and their RP-HPLC purifications were performed as previously described (Roubini et al. 1992).

10.2.2 In Vitro Activity: Binding, Adenylyl Cyclase and Intracellular Calcium

HEK-293 cells and hPTH1-receptor-expressing HEK-293/C-21 cells (about 400,000 receptors/cell) were cultured in D-MEM supplemented with 10% fetal bovine serum as described (Pines et al. 1994). For PTH1-receptor binding, HEK293/C-21 cells were subcultured in polylysine coated 24-wells plates and grown to confluency. Radioreceptor assays were carried out as previously described (Nakamoto et al. 1995) using ^{125}I-[Nle^{8,18},Tyr^{34}]PTH-(1–34) as radioligand. For adenylyl cyclase assays, HEK293/C-21 cells were subcultured in 24-wells plates and grown to near confluency. Determination of the activation of adenylyl cyclase by PTH analogues was performed as described (Nakamoto et al. 1995). Intracellular calcium release in fura-2/AM loaded HEK293/C-21 cells following stimulation by PTH(1–34) and other analogues was determined as described previously (Pines et al. 1996). Following their rigorous in vitro characterization some analogues were selected for in vivo characterization.

10.2.3 Photoaffinity Cross-Linking and Analysis

The photoaffinity cross-linking of the ^{125}I-benzophenone-substituted bPTH-(1–34) analogues were carried out as described (Zhou et al. 1997). In the studies described here, we have employed two specific benzophenone-containing analogues (Table 1). These analogues (Bpa-1 and K13) have biological activities indistinguishable from those of the parent peptide bPTH-(1–34) (Table 1).

Batches of SDS-PAGE purified radiolabeled hormone-receptor conjugate and fragments were prepared in small volumes (typically 10–20 µl) in 25 mM Tris HCl (pH 7.4), Triton X-100 (0.1% v/v), SDS (0.01% w/v).

Endo-N-glycosidase (Endo-F) digestions were carried out at 37°C for 24 h, according to the manufacturer's procedure. Lysyl endopeptidase C (Lys-C) digestions were performed by two 24 h treatments with 0.15 U (in 10 µl water) at 37°C. Cyanogen bromide (CNBr) digestions

Table 1. Biological activity of PTH photoaffinity ligands

Ligand	KD (M)	EC50 (adenylyl cyclase)	($[Ca^{2+}]_i$) (10^{-7} M)
PTH(1–34)	2×10^{-8}	0.8 nM	100 nM
K13	5×10^{-8}	1.2 nM	100 nM
Bpa-1	1×10^{-8}	2 nM	130 nM

were performed with 50 mg/ml CNBr in 70% formic acid at 37°C for 24 h in the dark. Samples were dried on Speed-Vac and dissolved in reducing sample buffer (Laemmli 1970) prior to electrophoresis.

10.2.4 Molecular Modeling

The molecular model of the hPTH1-receptor was generated using the topological arrangement of the TM helices of rhodopsin (Schertler et al. 1993). The locations of the TM portions of the hPTH1-receptor, assumed to be α-helices, were identified by a hydrophobicity profile of the PTH1-receptor sequence (Eisenberg et al. 1984). After inserting the identified helices of the PTH1-receptor into the rhodopsin template, the helices were rotated about their long axis, orienting the hydrophobic moment towards the membrane. Minor adjustments were made to the location of the helices following the substitution-table methodology reported by Donelly et al. (1994). A similar approach was used in an effort to develop the conformational preferences of the N-terminus and the intra- and extracellular loops of the receptor. The identified secondary structural features were incorporated into the developing molecular model. Molecular dynamics simulations and energy minimization were then performed.

10.3 Results and Discussion

10.3.1 Photoaffinity Ligand Characterization

Binding affinities for the two photoaffinity ligands (Bpa-1 and K13) were determined using competition with ^{125}I-[Nle^{8,18},Tyr^{34}]PTH-(1–34)NH$_2$ (^{125}IPTH-(1–34)) in hPTH1-receptor stably expressing HEK-293/C-21 cells. Agonist activity (stimulation of adenylyl cyclase activity and increases in intracellular calcium) were measured (Table 1). Both ligands have bioactivity (in vitro and in vivo) which is virtually indistinguishable from the parent peptide.

10.3.2 Photoaffinity Cross-Linking:
The Identification of Contact Domains

So far, the most powerful insights into the nature of the bimolecular interaction between hormone and receptor are yet to come. Historically, the manipulation of hormone structure in classical structure-activity relations has been performed in the absence of structural information of the receptor (by necessity), until the relatively recent cloning of the receptor (Jüppner et al. 1991; Schipani et al. 1993; Adams et al. 1995). However since that time, receptor mutagenesis studies have followed a similar pattern by focusing on a single component of the bimolecular interface, namely the receptor (Lee et al. 1995; Huang et al. 1996; Turner et al. 1996). As a result, both of these approaches share a common burden; the information generated by either is inferential at best.

The importance of a particular amino acid in the receptor can be identified, but the complementary interacting amino acid in the hormone cannot. A confounding issue is the fact that it cannot be known for certain why a particular amino acid substitution (in either the receptor or the hormone) impacts biological activity. Is it because the amino acid itself contains structural features which interact directly with the complementary molecule? Does the particular substitution cause a global conformational change in the molecule? Is the modification directly at an interacting site? These critical issues cannot be resolved by either approach alone. The only way to address these issues is via a direct, photoaffinity cross-linking approach, which provides experimentally based evidence for the location of specific cross-linking domains within the bimolecular interface or "hot spots."

In the studies shown here, using two biologically active benzophenone-containing PTH photoaffinity analogues (Table 1), we have been able to identify two specific sites of ligand-receptor interaction. The first direct identification of a hPTH-1 receptor/PTH cross-linking domain was achieved with the ligand K13 (Table 1). In these studies (Zhou et al. 1997), we examined the photoaffinity-labeled receptor-hormone conjugate by a series of chemical and enzymatic cleavages.

Photoaffinity cross-linking of ^{125}I-K13 to the hPTH1-receptor, stably expressed in HEK-293/C-21 cells, yielded a single diffuse band migrating at approximately 87 kDa (Fig. 1). This band is receptor-spe-

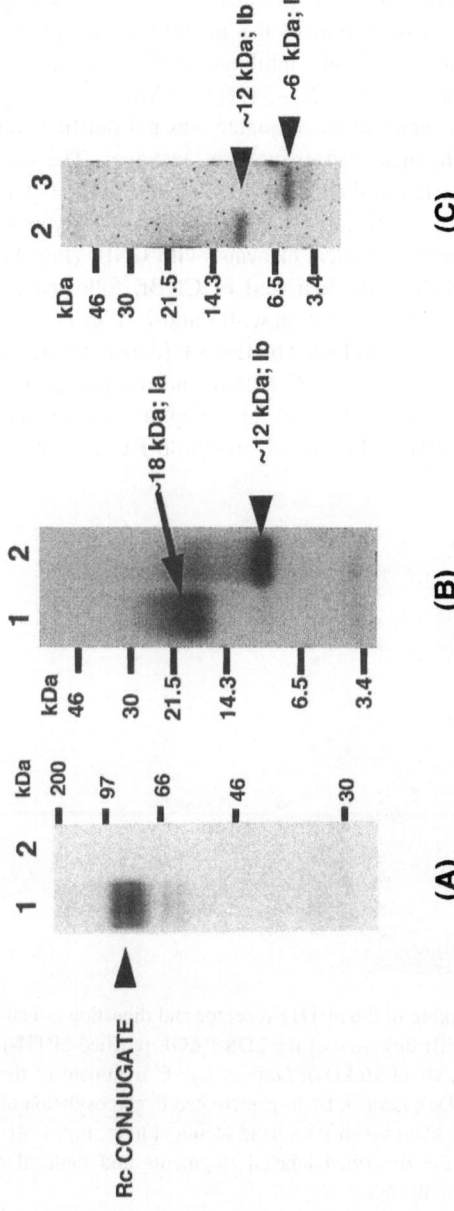

Fig. 1A–C. Photoaffinity cross-linking of the hPTH1-receptor and digestion modality I. **A** Photolabeling of C-21 cells expressing hPTH1-receptor by ^{125}I-K13 in the absence (*lane 1*) and presence (*lane 2*) of 10^{-6} M PTH(1–34). **B** The hPTH-1 receptor photoconjugate was first exhaustively digested by Lys-C (*lane 1*; Ia about18 kDa) and then deglycosylated with Endo-F (*lane 2*; Ib about 12 kDa). C The Endo-F-treated hPTH1-receptor-ligand fragment (*lane 1*) was completely digested with CNBr (*lane 2*; Ic about 6 kDa). *Arrows* show the position of the described labeled fragments and molecular weight markers are indicated at the *left* in each panel

cific, since it is not observed in similar experiments in receptor-lacking parental HEK-293 cells and formation of the ligand-receptor photo-cross-linked conjugate was completely inhibited in the presence of excess (1 μM) of unlabeled agonist PTH(1–34) (Fig. 1A).

The 87 kDa ^{125}I-K13-receptor photoconjugate was gel-purified and subjected to a series of chemical and enzymatic cleavages. The first digestion pathway (I) consisted of the enzymatic cleavage at the C-ter-minus of lysyl residues with Lys-C, followed by removal of N-glycosy-lation with Endo-F and finally chemical cleavage with CNBr (Fig. 1). The second digestion pathway (II) consisted of CNBr, followed by Lys-C treatment and finally deglycosylation with Endo-F (Fig. 2) and a third modality consisted of CNBr followed by Endo-F (data not shown).

Cleavage of the SDS-PAGE purified 87 kDa ligand-receptor photo-conjugate with Lys-C yielded a single band migrating at about 18 kDa (Ia) on 16.5% Tricine SDS-PAGE (Fig. 1B). Upon Endo-F treatment the

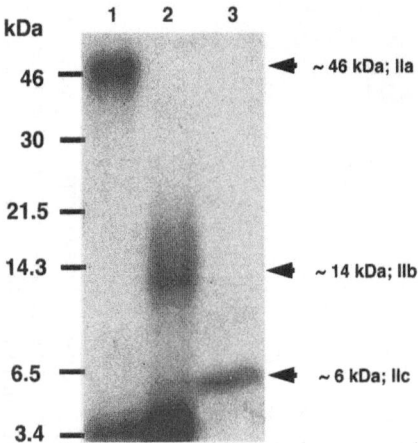

Fig. 2. Photoaffinity cross-linking of the hPTH1-receptor and digestion modal-ity II. *Lane 1*, Exhaustive CNBr digestion of the SDS-PAGE purified hPTH-1 receptor photoconjugate (IIa; about 46 kDa); *Lane 2*, Lys-C treatment of the 46 kDa band (IIb; about 14 kDa); *Lane 3*, Endo-F-mediated deglycosylation of the 14 kDa band (IIc; about 6 kDa) identifies a band identical to Ic, Fig. 1. *Ar-rows* show the position of the described labeled fragments and molecular weight markers are indicated at the *left*

molecular mass of this about18 kDa fragment (Ia) was reduced to approximately 12 kDa (Fig. 1B; Ib). Subsequent CNBr digestion produced a single approximately 6 kDa ligand-receptor fragment (Fig. 1C; Ic).

Similarly, cleavage of the SDS-PAGE purified 87 kDa ligand-receptor with CNBr produced a single, diffuse band of approximately 46 kDa (Fig. 2; IIa). Lys-C treatment of the about 46 kDa band released a fragment of approximately 14 kDa (Fig. 2; IIb). Subsequent deglycosylation of this diffuse band with Endo-F identified a single, sharply-resolved band of approximately 6 kDa (Fig. 2; IIc), identical to the 6 kDa fragment identified in pathway I (Fig. 1; Ic).

Following a similar strategy, using ^{125}I-Bpa-1-PTH-(1–34) and another series of chemical and enzymatic cleavages, we have identified another specific hormone-receptor contact domain (Fig. 3). The identification of the region of the receptor in contact with the principal activation domain of PTH-(1–34) (position 1–6) is of particular interest. The

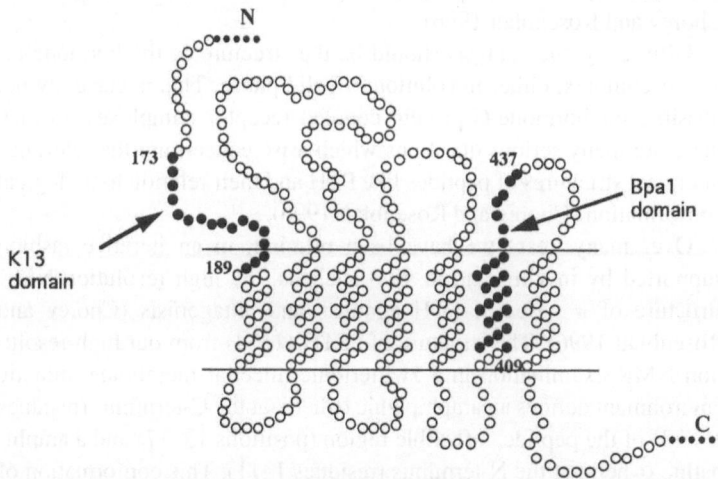

Fig. 3. The hPTH1-receptor showing the position of the identified contact domains. The smallest identified cross-linking domain for each benzophenone-containing PTH(1–34) ligand is shown as either the K13 domain or Bpa-1 domain. *Numbers* indicate the amino acid designation of the human receptor sequence; N-terminus is indicated by *N*, C-terminus by *C*

elucidation of the precise amino acid contact point will greatly enhance our understanding of the activation of the PTH-1 receptor.

10.3.3 Conformational Analysis: What the Hormone (and Receptor) Really Look Like

The elucidation of the three-dimensional structure of PTH is the major objective of structure-activity relations leading to the rational design of small molecule(s) with enhanced PTH-like pharmacological properties. With a large, flexible molecule like PTH all structural analysis is extremely challenging. In fact, the identification of a "structure" and the identification of a biologically relevant structure may be quite different. For a structure to be biologically relevant it must satisfy several stringent criteria: (1) it should explain structure-activity relations; (2) the conformation should resemble the conformation of analogues with similar biological activity; (3) the conformation should be distinct from analogues with different pharmacological profiles (Chorev et al. 1995; Chorev and Rosenblatt 1996).

Ultimately, the analysis should be the structure of the hormone-receptor complex, either in solution or solid phase. This is currently not possible for hormone-G-protein coupled receptor complexes. In fact, there are many serious questions which arise concerning the relevance of crystal structures of peptides like PTH and their relation to biological conformation (Chorev and Rosenblatt 1996).

Over many years we have been refining, in an iterative fashion supported by in vitro and in vivo analysis, the high resolution NMR structure of a series of PTH agonists and antagonists (Chorev and Rosenblatt 1996). The structure of hPTH-(1–34) from our high-resolution NMR examination in a zwitterionic micellar membrane mimetic environment defines an amphipathic α-helix at the C-terminus (residues 19–33) of the peptide, a flexible region (positions 13–17) and a amphipathic α-helix at the N-terminus (residues 1–11). This conformation of both PTH-(1–34) and PTHrP-(1–34) has been consistently observed under various experimental conditions.

The structure of the hPTH1-receptor obtained by our homology modeling and two-phase molecular dynamics indicates that the N-terminal receptor segment from Arg-179 to Arg-189 prefers the conformation of an extracellular amphipathic α-helix, whose axis is parallel to the membrane boundary and oriented away from the first TM helix bundle of the receptor. The identification of this novel structural feature within the hPTH1-receptor guided the docking of PTH-(1–34) in the molecular dynamic ligand-receptor interaction protocol.

10.3.4 Molecular Modeling: Putting It All Together

Our conformational studies of PTH suggest that the α-helices are an integral part of the bioactive conformation. Our direct identification of contact domains and our model of the PTH-1 receptor conformation enable us, for the first time, to generate a molecular simulation of the interaction of hormone and receptor.

In order to perform the simulation and develop the model, we optimized interactions between the amphipathic helix of the receptor (segment 179–186) just exterior of TM1 and that of hPTH-(1–34). Favorable ionic interactions between complementary charged side-chains positioning the helices in an antiparallel fashion were observed. This model positions residue 13 of hPTH(1–34) in close proximity to our recently identified cross-linked domain (Zhou et al. 1997). Stabilizing the conformation of the docking model around this direct interaction, we looked for interactions between the N-terminus PTH-(1–34) and the helix bundle of the receptor. In TM6 the two methionines (Met-414 and Met-425) were identified as putative interaction points with the N-terminal amino acid of PTH-(1–34). Throughout the docking procedure, deviations from the experimentally determined structure of hPTH(1–34) were not allowed.

The model enables us to make predictions regarding the identity of receptor points which maybe important for the interaction between ligand and receptor. Using our specialized photoaffinity cross-linking approach we can and are testing these predictions directly.

10.4 Summary

10.4.1 The Future: Using and Developing the Model

The combination of direct physical identification of cross-linking points, structural analysis of hormone and receptor along with site-directed mutagenesis enables the refinement of our model. We are in the unique position of being able to develop a molecular model of hormone-receptor activation which is supported by direct experimental evidence.

Increased understanding of the nature of specific bimolecular interactions between PTH and the hPTH-1 receptor should enable analogue design in a manner not previously possible. The impact of such studies on rational drug design for small molecule PTH mimetics with anabolic effects on bone will be spectacular. Since technology inevitably exceeds what is currently perceived to be possible, the only certainty about the future of PTH-based drug discovery is that it will be even more exciting.

Acknowledgements. I would like to thank my friends and colleagues who performed the experiments described here. In particular, Alessandro Bisello, Alice T. Zhou, Amy E. Adams, Chizu Nakamoto, Maria Pelligrini, and Dale Mierke. I would also like to thank Michael Chorev and Michael Rosenblatt for their continued support of my efforts.

References

Abou-Samra AB, Jüppner H, Westerberg D, Potts Jr JT, Segre GV (1989) Parathyroid hormone causes translocation of protein kinase-C from cytosol to membranes in rat osteosarcoma cells and murine T-lymphoma cells. Endocrinology 124:1107–1113

Adams AE, Pines M, Nakamoto C, Behar V, Yang QM, Besalle R, Chorev M, Rosenblatt M, Suva LJ (1995) Probing the bimolecular interactions of parathyroid hormone and the parathyroid hormone/parathyroid hormone-related protein receptor. 2. Cloning, characterization and photoaffinity labeling of the recombinant human receptor. Biochemistry 34:10553–10559

Bisello A, Mierke DF, Pelligrini M, Rosenblatt M, Suva LJ, Chorev M (1998) Parathyroid hormone-receptor interactions identified directly by photocross-linking and molecular modeling studies. J Biol Chem (in press)

Chorev M, Behar V, Yang Q, Rosenblatt M, Mammi S, Maretto S, Pellegrini M, Peggion E (1995) Conformation of parathyroid hormone antagonists by CD, NMR, and molecular dynamics simulations. Biopolymers 36:485–495

Chorev M and Rosenblatt M (1994) In: Bilezikian JP, Marcus R, Levine M (eds) The parathyroids: basic and clinical concepts . Raven, New York, pp 139–156

Chorev M and Rosenblatt M (1996) In: Bilezikian JP, Raisz LG, Rodan GA (eds) Principles of bone biology. Academic, San Diego, pp 305–323

Dempster DW, Cosman F, Parisien M, Shen V, Lindsay R (1993) Anabolic actions of parathyroid hormone on bone. Endocrine Rev 14:690–709

Donelly D, Overington JP, Blundell TL (1994) The prediction and orientation of alpha-helices from sequence alignments: the combined use of environment-dependent substitution tables, Fourier transform methods and helix capping rules. Prot Engin 7:645–653

Eisenberg D, Weiss RM, Terwilliger TC (1984) The hydrophobic moment detects periodicity in protein hydrophobicity. Proc Natl Acad Sci USA 81:140–144

Gardella TJ, Jüppner H, Wilson AK, Keutmann HT, Abou-Samra AB, Segre GV, Bringhurst R, Potts Jr. JT, Nussbaum SR, Kronenberg HM (1994) Determinants of [Arg2]PTH-(1–34) binding and signaling in the transmembrane region of the parathyroid hormone receptor. Endocrinology 135:1186–1194

Gardella TJ, Luck MD, Fan MH, Lee CW (1996) Transmembrane residues of the parathyroid hormone (PTH)/PTH-related peptide receptor that specifically affect binding and signaling by agonist ligands. J Biol Chem 271:12820–12825

Huang Z, Chen Y, Pratt S, Chen TH, Bambino T, Nissenson RA, Shoback DM (1996) The N-terminal region of the third intracellular loop of the parathyroid hormone (PTH)/PTH-related peptide receptor is critical for coupling to cAMP and inositol phosphate/Ca^{2+} signal transduction pathways. J Biol Chem 271:33382–33389

Iida-Klein A, Varlotta V, Hahn TJ (1989) Protein kinase C activity in UMR 106-01 cells: effects of parathyroid hormone and insulin. J Bone Miner Res 4:767–774

Jüppner H, Abou-Samra AB, Freeman MW, Kong XF, Schipani E, Richards J, Kolakowski LF, Hock J, Potts Jr. JT, Kronenberg HM, Segre GV (1991) A G protein-linked receptor for parathyroid hormone and parathyroid hormone-related peptide. Science 254:1024–1026

Laemmli UK (1970) Cleavage of structural proteins during the assembly of the head of bacteriophage T4. Nature 227:680–685

Lee C, Luck MD, Jüppner H, Potts JT Jr, Kronenberg HM, Gardella TJ (1995) Homolog-scanning mutagenesis of the parathyroid hormone (PTH) receptor reveals PTH-(1–34) binding determinants in the third extracellular loop. Mol Endocrinol 9:1269–1278

Nakamoto C, Behar V, Chin K, Adams AE, Suva LJ, Rosenblatt M, Chorev M (1995) Probing the bimolecular interactions of parathyroid hormone (PTH) and the human PTH/PTHrP receptor. I. Design, synthesis and characterization of photoreactive benzophenone-containing analogues of PTH. Biochemistry 34:10546–10552

Partridge NC, Kemp BE, Veroni MC, Martin TJ (1981) Activation of adenosine 3', 5'-monophosphate-dependent protein kinase in normal and malignant bone cells by parathyroid hormone, prostaglandin E_2 and prostacyclin. Endocrinology 108:220–225

Pellegrini M, Bisello A, Rosenblatt M, Chorev M, Mierke DF (1997) Conformational studies of RS-66271, an analogue of parathyroid hormone-related protein with pronounced bone anabolic activity J Med Chem 40: 3025–3031.

Pines M, Adams AE, Stueckle S, Bessalle R, Rashti-Behar V, Chorev M, Rosenblatt M, Suva LJ (1994) Generation and characterization of human kidney cell lines stably expressing recombinant human PTH/PTHrP receptor: Lack of interaction with a C-terminal human PTH peptide. Endocrinology 135:1713–1716

Pines M, Fukayama S, Costas K, Meurer EM, Goldsmith PK, Xu X, Muallem S, Behar V, Chorev M, Rosenblatt M, Tashjian Jr. AH, Suva LJ (1996) Inositol 1, 4, 5 trisphosphate-dependent Ca^{2+} signaling by the recombinant human PTH/PTHrP receptor stably expressed in a human kidney cell line. Bone 18:381–389

Roubini E, Doung LT, Gibbons SW, Leu CT, Caulfield MP, Chorev M, Rosenblatt M (1992) Synthesis of fully active biotinylated analogueues of parathyroid hormone and parathyroid hormone-related protein as tools for the characterization of parathyroid hormone receptors. Biochemistry 31:4026–4033

Schertler GF, Villa C, Henderson R (1993) Projection structure of rhodopsin. Nature 362:770–772

Schipani E, Karga H, Karapalis AC, Potts Jr. JT, Kronenberg HM, Segre GV, Abou-Samra AB, Jüppner H (1993) Identical complementary deoxyribonucleic acids encode a human renal and bone parathyroid hormone (PTH)/PTH-related peptide receptor. Endocrinology 132:2157–2165

Turner PR, Bambino T, Nissenson RA (1996) Mutations of neighboring polar residues on the second transmembrane helix disrupt signaling by the parathyroid hormone receptor. Mol Endocrinol 10:132–139.

Usdin TB, Gruber C, Bonner TI (1995) Identification and functional characterization of a receptor selectively recognizing parathyroid hormone, the PTH2 receptor. J Biol Chem 270:15455–1545

Zhou AT, Bessalle R, Bisello A, Nakamoto C, Rosenblatt M, Suva LJ, Chorev M (1997) Direct mapping of an agonist-binding domain within the parathyroid hormone/parathyroid hormone-related protein receptor by photoaffinity labeling. Proc Natl Acad Sci USA 94:3644–3649

11 Bone Morphogenetic Proteins: Role, Modes of Action and Potential Significance in Osteoporosis

P.J. Marie

11.1 Introduction

The osteoblast differentiation pathway is characterized by the commitment of osteoprogenitor cells from mesenchymal stem cells, the progressive differentiation of pre-osteoblasts associated with the expression of genes of the osteoblast phenotype, and the synthesis, deposition and mineralization of bone matrix by post-mitotic mature osteoblasts. This differentiation process is controlled by a variety of local factors acting on skeletal cell proliferation and differentiation to initiate and promote bone matrix formation during skeletal development, formation and repair (Aubin et al. 1993; Baylink et al. 1993; Mundy 1993; Marie and DeVernejoul 1993; Marie et al. 1994).

Among the various local factors known to stimulate bone formation, a family of proteins play a major role in the differentiation of skeletal mesenchymal cells. More than 30 years ago, Urist (1965) showed that protein extracts from demineralized bone matrix implanted at nonskeletal sites induce the de novo formation of cartilage and bone tissue in vivo. These bone morphogenetic proteins (BMPs) present in the bone matrix induce a cascade of events that recapitulates the process of endochondral bone formation (Urist et al. 1983; Reddi and Cunningham 1993). Subsequent extensive research was aimed at better understanding the mode of action of BMPs on skeletal development and bone formation. BMPs were shown to initiate the commitment of mesenchymal cells, to promote the differentiation of committed cells into skeletal cells, and to stimulate the expression of differentiation markers in these cells (Wozney 1992).

The research developed in my laboratory for the last 10 years has focused on the study of the biology and pathology of bone formation at the tissue, cellular and molecular levels. We have been particularly interested in human osteoblast biology and pathology (Marie and De-Vernejoul 1993, Marie 1995), the effects of anabolic agents capable of increasing osteoblast cell proliferation and/or differentiation, and promoting bone formation and bone mass in experimental models (Marie 1997a). We recently developed new models of human osteoblastic cells which allowed us to obtain new information on the mode of actions of BMPs in human differentiation. This review summarizes our current knowledge of the modes of action of BMPs as well as more recent developments regarding the pleiotropic effects and modes of action of BMPs on cells of the osteoblastic lineage, with a particular emphasis on human osteoblasts and the potential role of BMPs in the treatment of osteoporosis.

11.2 The Bone Morphogenetic Protein Family

Purification of proteins with BMP activity (Sampath et al. 1987; Wang et al. 1988, Luyten et al. 1989) led to the cloning of seven proteins, BMPs 1–7 (Wozney et al. 1988). Sequence similarities have been found in several members of the BMP subfamily belonging to the TGF-β superfamily. BMPs have been classified into subgroups (BMP-2 and BMP-4 form one group; BMPs 5–8 form another group) on the basis of sequence homology in the cysteine portion of the mature region of the protein. BMPs are highly conserved proteins that are related to several developmental regulatory genes (decapentaplegic (*dpp*) and *Vgr*/60Agene in *Drosophila*, vegetal (*Vg*-1) in *Xenopus*). BMP-2 and -4 appear to be the mammalian homologues of *dpp* in Drosophila, whereas BMP-5, -7, and -8 have significant homology with *Vgr*/60A (Table 1) (Kingsley 1994). Although BMPs show some structural analogy, they exert distinct biological actions on skeletal development, induction of mesenchymal cell differentiation and promotion of cartilage and bone formation.

11.3 Role of Bone Morphogenetic Proteins in Skeletal Development

Some BMPs and related genes appear to play a major role during skeletal development and patterning (Reddi and Cunningham 1993; Hogan 1996; Monsoro-Burq et al. 1996; Rosen et al. 1996). For example, BMPs are differentially expressed at various stages in different tissues during embryonic formation (Lyons et al. 1989; Ozkaynak et al. 1992). During limb and craniofacial development, BMP-2, -3, -4, -5 and

Table 1. Members of the bone morphogenetic protein (BMP) subfamily

dpp subfamily	60A subfamily	Others
BMP-2	BMP-5	BMP-3
BMP-4	BMP-6	Vg1
dpp	BMP-7/OP1	GDF1
	BMP-8/OP2	

BMPs 2–8 are involved in skeletal cell differentiation in mammals.

-7 transcripts are distinctly localized in regions surrounding cartilage or mesenchymal condensation (Lyons et al. 1990; Vukicevic et al. 1990a, Francis-West et al. 1995) and during tooth formation (Vainio et al. 1993). The distinct localization of BMP-2 and BMP-4 during embryogenesis suggests that BMPs may play different roles in determining skeletal patterning during limb formation (Francis et al. 1994). This is emphasized by the finding that overexpression of BMP-2 and -4 alters the size and shape of developing skeletal elements in the chick limb (Duprez et al. 1996). BMP-2 and -4 deletions in mice are lethal while BMP-7-deficient mice show mesodermal and skeletal defects associated with polydactyly in hindlimbs (Luo et al. 1995). It also appears that BMP-4 and BMP-7 are implicated in mesenchymal development in connection with other developmental genes (Lyons et al. 1990; Niswander et al. 1994; Tickle 1994). For example, growth and differentiation factors (GDFs) that are closely related to the BMP subfamily are also involved in skeletal morphogenesis. Disruption of the GDF-5 gene induces alteration of the limb skeletal pattern (Storm et al. 1994). In the BMP-5 mutant mouse, bone and cartilage embryonic development are defective, suggesting that BMP-5 is involved in postnatal bone formation (King et al. 1994). Moreover, BMP-5 and GDF-5 control the development of different cartilage elements (Kingsley 1994). Normal morphogenesis appears therefore to imply the actions of specific members of the TGF-β superfamily.

BMPs were found to play an important role in limb development. They are involved in the establishment of the anteroposterior axis of the limb, and their expression is related to the signaling cascade involved in limb patterning (Francis-West et al. 1995). In the embryo, BMPs act as inductive signals between tissue layers and regulate the expression of several genes. Overexpression or deletion studies showed that members of *Hox*d genes regulate skeletal formation (Dollé et al. 1993; Hogan 1996). BMPs activate *Hox*d genes (Duprez et al. 1996), suggesting that BMPs play a role in patterning of the limb through *Hox*d gene expression (Hofmann et al. 1996). By contrast, BMP-2 and -7 are potent apoptotic inducers of undifferentiated limb mesoderm (Macias et al. 1997), and part of this effect may be mediated by the induction of *Msx*-1 and *Msx*-2, transcription factors belonging to the basic helix-loop-helix (bHLH) family. BMP-4 was shown to modulate the expression of Msx-1 and Msx-2 in the embryo (Wand and Sassoon 1995; Marazzi et al. 1997) and this factor mediates the program of cell death induced by BMP-4 (Marazzi et al. 1997). Introduction of a dominant-negative BMP receptor results in decreased interdigital cell death and *Msx*-2 transcripts in the embryonic limb, indicating that BMP signaling mediates cell death (Zou and Niswander 1996). *Msx* genes are also involved in the BMP-4-induced differentiation of vertebral cartilage from the somitic mesenchyme (Monsoro-Burq et al. 1996), and BMP-4 produced by early dental epithelium regulates tooth-specific gene expression in the dental mesenchyme along with the expression of Msx-2 (Thesleff 1995). It is noteworthy that *Msx*-1 expression is decreased in BMP-7-deficient mice (Hofmann et al. 1996), further stressing the role of *Msx* genes in BMP-induced skeletal development.

Recent data indicate that hedgehog genes and BMPs are coexpressed at diverse sites in the embryo. Hedgehog encodes a protein that regulates embryonic segmentation and patterning during fly development. Vertebrate homologues (sonic hedgehog and Indian hedgehog) have multiple functions during skeletal formation in limbs and vertebrae (Bitgood and McMahon 1995). Moreover, BMP-2 is activated by sonic hedgehog (Shh) (Laufer et al. 1994). Patterning of the vertebrate limb may thus be directed by signaling molecules induced by Shh acting together with BMP-2 (Rodriguez et al. 1996).

11.4 Effects of Bone Morphogenetic Proteins on Mesenchymal Cell Differentiation

The differentiation of mesenchymal stem cells toward osteoblasts, chondroblasts, adipocytes or myocytes is under the control of BMPs acting at different steps (summarized in Reddi 1995; Yamaguchi, 1995) (Fig. 1).

11.4.1 Myoblastic Cells

Bone morphogenetic proteins have been shown to negatively regulate muscle cell differentiation in vitro (Yamaguchi et al. 1991). For example, BMP-2 converts the myoblastic cell line C2C12 cells into osteoblasts (Katagiri et al. 1994). In fact, BMPs inhibit the expression of myogenic determination genes (myogenin, *Myo*D, herculin, and *myf*-5)

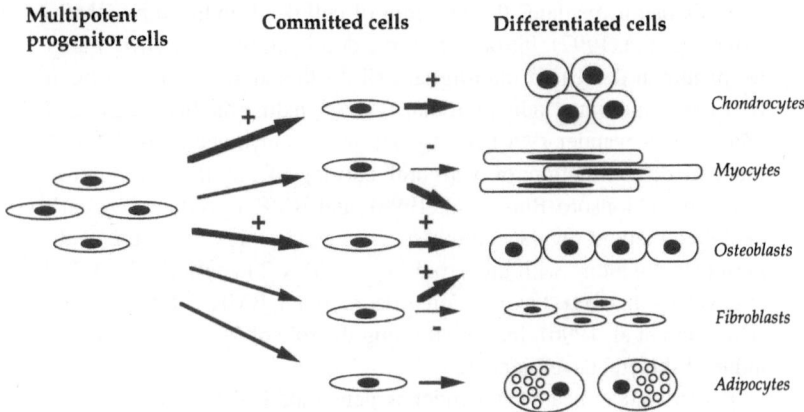

Fig. 1. Effects of bone morphogenetic proteins (BMPs) on mesenchymal cell differentiation. BMPs induce (+) the differentiation of multipotent progenitor cells to committed cells, promote (+) the differentiation of committed cells to chondroblasts or osteoblasts, and inhibit (−) formation of myocytes and fibroblasts

in myoblastic cells (Murray et al. 1993). In addition, BMP-2 orientates the calvaria-derived C26 cell line, which is able to differentiate into osteoblasts, adipocytes or myocytes, towards osteoblastic cells expressing alkaline phosphatase (ALP) activity, osteocalcin (OC) synthesis and parathyroid hormone (PTH)-dependent cAMP production, but does not affect the more mature osteoblast-like cell line C20, suggesting that BMP-2 has different effects depending on the stage of cell differentiation (Yamaguchi et al. 1991). Continuous treatment with BMP-2 is required for the conversion from muscle cells to osteoblasts, suggesting that BMP-2 induces the differentiation of committed osteoprogenitors to more mature osteoblasts (Yamaguchi et al. 1991). Other BMPs may however have different effects since BMP-12 and -13 inhibit terminal differentiation of myoblasts but do not induce their differentiation into osteoblasts (Inada et al. 1996).

11.4.2 Cartilaginous Cells

Bone morphogenetic proteins may initiate the differentiation of multipotent mesenchymal progenitor cell lines to the chondrogenic or osteogenic lineages. For example, BMP-2 and -4 induce the differentiation of C3H10T1/2CL8 mouse embryonic cells to osteoblastic cells (Katagiri et al. 1990; Ahrens et al. 1993). By contrast, BMPs can induce the differentiation of early skeletal progenitor cells into chondroblastic cells. For example, BMP-3 and -4 induce chondrogenesis in chick limb bud mesenchymal cells (Carrington et al. 1991), whereas BMP-2 induces differentiation of the pluripotent undifferentiated C3H10T1/2 cell line into chondroblasts (Ahrens et al. 1993; Wang et al. 1993). Although BMP-2 or -4 or BMP-7 can induce the differentiation of newborn rat calvaria cells into chondroblasts (Asahina et al. 1993), BMPs are not able to reverse the osteoblast to the chondrocyte phenotype (Asahina et al. 1993; Komaki et al. 1996).

BMPs can act on already differentiated chondroblasts to maintain the cartilage phenotype (Rosen et al. 1994). For example, BMP-2 and-3 stimulate the chondroblast phenotype in rat and rabbit articular chondrocytes (Vukicevic et al. 1989; Hiraki et al. 1991) and promote expression of the cartilage phenotype in dedifferentiated rabbit articular chondrocytes (Harrison et al. 1991). In addition, BMP-3 and 4 enhance the cartilaginous phenotype in rabbit chondrocytes (Luyten et al. 1992), and BMP-7 supports maturation of embryonic chick sternal chondrocytes (Chen et al. 1995) and stimulates chondrogenesis in long bone rudiments (Dieudonne et al. 1994) and human articular chondrocytes (Flechtenmacher et al. 1996). It is noteworthy that BMP-2 first initiates the differentiation of limb bud cells into cartilage cells and then induces these cells to express bone matrix proteins, indicating that BMP-2 can induce skeletal cell differentiation in a sequential manner (Rosen et al. 1994). Other observations suggest that TGF-β3 and BMP-2 may be acting in a sequential manner to regulate chick limb mesenchymal cells through the different stages of cartilage differentiation (Roark and Greer 1994).

11.4.3 Marrow Stromal Cells

Bone marrow stroma cells contain a small fraction of stem cells that can differentiate to osteoblasts under appropriate stimulation (Owen 1985). Thies et al. (1992) first showed that BMP-2 increases ALP activity, OC synthesis and PTH sensitivity, as indicated by cAMP production, in the clonal W-20-17 cell line derived from mouse bone marrow stroma. BMP-2 and -3 were also shown to increase ALP activity in rat clonal stromal cell lines (Benayahu et al. 1994). In the bone marrow stromal cell line ST2, BMP-2 stimulates ALP activity, OC production, and PTH-dependent cAMP production, and BMP-2 and -4 are more potent than BMP-6 in inducing the osteoblast phenotype (Yamaguchi et al. 1996), indicating that BMPs are able to induce the differentiation of bone marrow stromal cells into osteoblasts (Fig. 1).

BMPs also induce osteoblast differentiation in stromal cells in humans. In human bone marrow stromal cells, BMP-3 inhibits cell proliferation and stimulates ALP activity and OC production (Amedée et al. 1994). Using a new model of human stromal cell differentiation in long-term culture (Fromigué et al. 1997), we recently found that BMP-2 promotes human marrow stromal cell differentiation into osteoblasts and counteracts the action of TGF-β2 (Fromigué et al. 1998). In this model, TGF-β2 increases DNA content and stimulates type I collagen synthesis, but inhibits ALP activity and mRNA levels, OC production, and mineralization of the matrix formed by human bone marrow stromal cells. By contrast, BMP-2 increases ALP activity and mRNA levels, OC levels and calcium deposition in the extracellular matrix without affecting type I collagen synthesis and mRNA levels, showing that BMP-2 and TGF-β2 have differential effects on these cells (Fromigué et al. 1998) (Fig. 2). Other factors may act in concert with BMPs on marrow stromal cell differentiation. For example, we recently found that 1,25(OH)$_2$ vitamin D and the synthetic glucocorticoid dexamethasone act in a sequential manner with BMP-2 and TGF-β2 to regulate human bone marrow stromal cell proliferation and to promote the osteoblast phenotype in human marrow stromal cells (Fromigué et al. 1998) (Fig. 2). Interestingly, BMP-2 was also found to cooperate in enhancing the osteogenic potency of basic fibroblast growth factor (bFGF) in rat marrow stromal cell cultures (Hanada et al. 1997). These results indicate

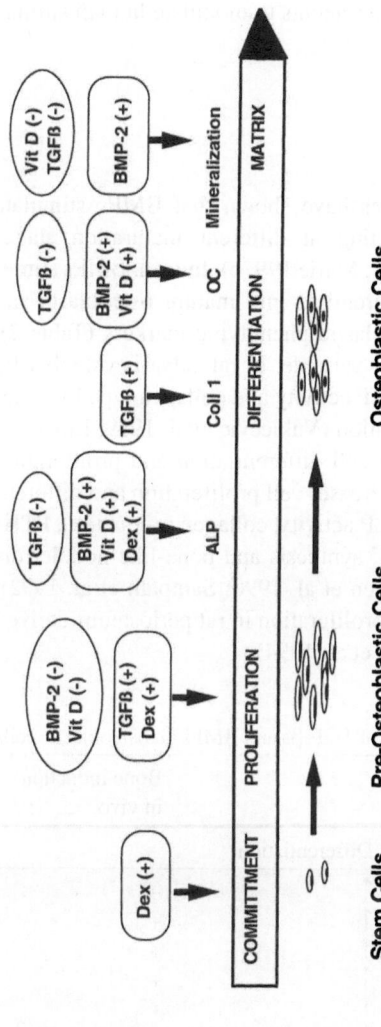

Fig. 2. Effects of local and hormonal factors on the proliferation and differentiation of human stromal cells to the osteoblast phnotype. (From Fromigué et al. 1997, 1998)

that BMPs act in concert with other agents to modulate human stromal
cell differentiation (Fig. 2).

11.4.4 Osteoblastic Cells

A variety of experimental studies have shown that BMPs stimulate
osteoblast differentiation by acting at different maturation stages
(Mundy 1993; Linkhart et al. 1996; Marie 1997b). Indeed, BMPs stimulate preosteoblastic cells to differentiate into mature osteoblasts and
increase the expression of osteoblastic phenotypic markers (Table 2).
The effects of BMPs are however variable. In rat calvaria osteoblastic
cell cultures, BMP-3 increases ALP activity and collagen type I synthesis without affecting cell proliferation (Vukicevic et al. 1989; Luyten et
al. 1992). BMP-4 promotes both cell differentiation and proliferation
(Chen et al. 1991), and BMP-7 increases cell proliferation and differentiation, as shown by increased ALP activity, collagen production, PTH-
stimulated cAMP production, OC synthesis and bone-like nodule formation in the same model (Knutsen et al. 1990; Sampath et al. 1992).
By contrast, BMP-2 inhibits cell proliferation in rat periosteum-derived
osteoblastic cell cultures (Iwasaki et al. 1994).

Table 2. Comparison between effects of TGF-βs and BMPs on osteoblastic cells

Factor	Osteoblastic cells		Bone induction in vivo
	Proliferation	Differentiation	
TGF-β	↑	↑	–
BMP-2	–↓[a]	↑	+
BMP-3	–↓[a]	↑	+
BMP-4	↑	↑	+
BMP-5	?	↑	+
BMP-6	?	↑	+
BMP-7/OP1	↑↓[a]	↑	+
BMP-8/OP2	?	↑	+

TGF, transforming growth factor; BMPs, bone morphogenetic proteins; ↑, increase; ↓, decrease; –, no effect; ?, unknown; +, positive effect.
[a]Variable effect depending on the cell type studied.

We recently developed and characterized a new model allowing us to delineate the mechanisms of action of agents that control the proliferation and differentiation of human calvarial cells during membranous osteogenesis. The phenotypic characteristics of human calvarial cells were determined by immunocytochemical, biochemical and molecular methods (De Pollak et al. 1996, 1997; Lomri et al. 1997; Debiais et al. 1998) (Fig. 3). In basal culture conditions, all cells express type I collagen, a fraction of cells express ALP, osteopontin, osteonectin, PTH-related protein (PTHrP) and PTHrP receptor and OC, showing that the cells express early markers of the osteoblast phenotype. In addition, the cells express constitutively FGF-2, FGF receptor (FGFR)-1 and FGFR-2 and produce TGF-β2 (Debiais et al. 1998). These phenotypic characteristics of human calvarial cells in culture are similar to those found in situ in calvarial samples (De Pollak et al. 1997), as evaluated by immunohistochemistry and biochemical methods, indicating that the cell behavior in culture is relevant to the in vivo situation. We then determined the temporal changes in osteoblast markers during human calvarial cell differentiation and osteogenesis in long-term culture in the presence of ascorbic acid and phosphate. In this model, human calvarial cells first proliferate, express high ALP activity and synthesize type 1 collagen, then ALP activity and collagen synthesis decrease progressively with time, the extracellular matrix accumulates and becomes mineralized, and calcium incorporation into the extracellular matrix increases with time in culture (Fig. 3). As the effect of BMPs in the induction of human osteogenesis is not known, we studied the cellular osteoblastic response to BMP-2 in this model. To determine at what stage BMP-2 is acting, the cells were treated with BMP-2 at different stages of differentiation in long-term culture (Fig. 3). We found that recombinant human BMP-2 (rhBMP-2) affects human calvarial cells differently depending on the stage of cell maturation. Indeed, BMP-2 was more potent on less mature cells in inducing matrix mineralization than on more mature cells (Table 3), and the promotion of matrix mineralization induced by BMP-2 was independent of the changes in cell proliferation and bone matrix protein synthesis (Hay et al. 1998). We are currently investigating changes in the early genes that are involved in this effect on immature human calvarial cells. It is noteworthy that the effects of BMP-7 on bone matrix expression by fetal rat calvarial cells are also differentiation stage-specific (Li et al. 1996).

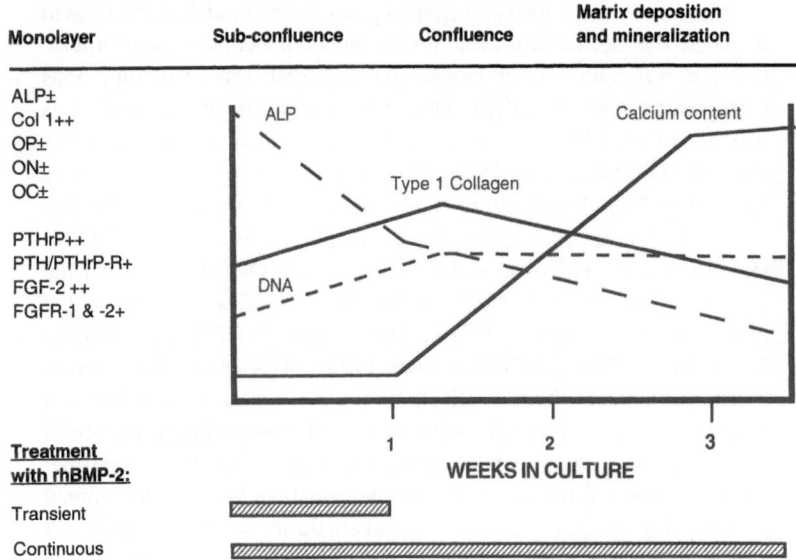

Fig. 3. Phenotype of human calvaria cells and temporal changes in differentiation markers during in vitro osteogenesis: a model to study the effects on bone morphogenetic proteins (BMPs) on human calvaria osteogenesis. (From De Pollack et al. 1996; Lomri et al. 1997; Hay et al. 1998; Debiais and Marie 1998)

BMPs may affect osteoblastic cells in several other ways. For example, BMP-2 was found to regulate collagenase-3 in fetal rat calvariae (Varghese and Canalis 1997) and to stimulate the migration of human osteoblasts, which may play a role during bone healing and remodeling (Lind et al. 1996). BMP-2 not only affects matrix proteins, but also cell surface expression of particular integrins in osteosarcoma cells and chondrocytes, suggesting that BMP-2 is a regulator of cell adhesion (Nissinen et al. 1997). It is of interest that, although BMPs have been described as homodimers (Rosen and Thies 1992), coexpression of BMP-2 with BMP-7 may yield BMP-2/BMP-7 heterodimers which are more potent than homodimers in inducing osteoblast differentiation in vitro and bone formation in vivo (Israel et al 1996). Finally, it is also of interest that, in addition to osteoblasts, BMPs regulate cell differentia-

Table 3. Differential effects of BMP-2 on human calvaria cells at different stages of differentiation. (From Hay et al. 1997)

Stage of differentiation	DNA synthesis	ALP expression and activity	Type 1 collagen synthesis	Osteocalcin expression and synthesis	Matrix mineralization
Immature cells	–	+	–	0	++
Mature cells	–	+	–	+	+

ALP, alkaline phosphatase; –, inhibition; + and ++, small and marked stimulation, respectively; 0: no effect.

tion in other skeletal tissues since BMP-2 and -4 induce the differentiation of dental pulp cells into preodontoblasts (Nakashima et al. 1994).

Fetal rat calvarial cells express BMP-2, -3, -4 and -6 (Harris et al
1994; Chen et al 1995), and BMP-2 enhances BMP-3 and -4 gene
expression during in vitro osteogenesis induced by fetal rat calvaria cells
(Chen et al. 1997). In this model, the expression of BMP-2 coincides
with the induction of ALP, OC and the formation of nodules (Harris et al
1994), and BMP-2, -4 and -6 increase the expression of ALP and OC
and enhance the osteogenic differentiation process (Ghosh-Choudhury
et al. 1994; Hughes et al. 1995), suggesting that BMPs may act as
autocrine regulatory factors. Interestingly, BMP-2 increases the
osteoblast phenotype and accelerates nodule formation in osteoblastic
cells cloned from the calvaria of transgenic mice that express SV40 T
antigen (Ag) under the control of the BMP-2 promoter. In this system,
exogenous BMP-2 regulates the expression of BMP-2 (Ghosh-Choudhury et al. 1996).

In immortalized cells and osteosarcoma cells, BMPs generally promote expression of the osteoblast phenotype, although they have divergent effects on cell proliferation (Table 2). For example, BMP-2 promotes cell differentiation but does not affect cell growth in immortalized
HOBIT human osteoblast-like cells (Zheng et al. 1994), whereas BMP-
2 and -3 inhibit cell proliferation in mouse calvaria-derived MC3T3-E1
cells (Vukicevic et al. 1990b). In rat osteosarcoma 17/2.8 cells or
MC3T3-E1 cells, BMP-7 inhibits cell proliferation (Maliakal et al.
1994; Rudkin et al. 1996) but stimulates cell proliferation in human
osteosarcoma cells, an effect mediated by the increased production of
insulin-like growth factors (IGFs) (Knutsen et al. 1995). Thus, BMPs
have divergent effects according to the cell type or stage of maturation.

Several interactions of BMPs with hormones have been described.
The effects of BMPs on skeletal cells are modulated by glucocorticoids
which potentiate the effects of BMP-2, -4 and -6 on the differentiation
of rat marrow stromal cell and rat calvarial cells (Rickard et al. 1994;
Boden et al. 1996). Vitamin D and estradiol also synergize with BMP-2
to stimulate ALP activity in MC3T3-E1 cells (Takuwa et al. 1991).
BMP-2 has both synergistic and antagonistic effects with retinoic acid
on osteoblastic markers in rat calvaria cell lines (Ogata et al. 1994), and
retinoic acid decreases BMP-2 and-4 expression in multipotent mesenchymal C26 and 10T1/2 cells (Gazit et al. 1993), indicating that

several hormonal factors modulate the effects of BMPs on osteoblastic cells.

Although belonging to the same TGF-β superfamily, BMPs and TGF-β often have opposite effects on skeletal cells. For example, TGF-β2 induces cell proliferation in normal rat calvarial cells (Machwate et al. 1995) and human osteoblastic cells (Lomri and Marie 1990), whereas BMP-2 has mainly inhibitory effects on osteoblastic cell growth, as discussed above. In a variety of cellular models, TGF-β increases collagen type I and reduces ALP and OC expression (Rodan and Noda 1991), whereas BMPs have opposite effects (Chen et al. 1991; Harrison et al. 1991; Hiraki et al. 1991; Sampath et al. 1992; Asahina et al. 1993; Zhou et al. 1993; Nakashima et al. 1994). Moreover, several interactions were found between TGF-β and BMPs. For example, we recently reported in human marrow stromal cells that TGF-β2 attenuates the stimulatory effect of BMP-2 on ALP activity and mRNA levels and on OC production, whereas BMP-2 reduces TGF-β2-enhanced DNA synthesis and type I collagen synthesis (Fromigué et al. 1998). We also found that a transient treatment with BMP-2 abolishes the TGF-β2 effect on ALP activity, whereas a transient treatment with TGF-β2 does not influence the subsequent BMP-2 action on human marrow stromal cell differentiation, stressing the complex interactions between TGF-β and BMP. TGF-β and BMPs have also been found to interact on their own production. BMP-2 increases TGF-β expression in osteoblastic cells (Dallas et al. 1994; Zheng et al. 1994), whereas TGF-β increases BMP-2 expression in rat calvaria osteoblasts (Harris et al. 1994). It was recently reported, however, that TGF-β1 and BMP-7 in vivo may synergize to initiate bone formation in primates (Ripamonti et al. 1997). It is likely that skeletal cells produce different members of the TGF-β superfamily at distinct stages of differentiation, and these factors may act in a sequential manner along the osteoblast/chondroblast differentiation pathway to regulate cell proliferation and differentiation (Fig. 2).

Finally, it is noteworthy that the effects of BMPs on skeletal cells may be mediated in part by other factors. For example, BMP-2 increases IGF-I and -II expression in rat calvarial osteoblasts (Canalis and Gabbitas 1994) and increases TGF-β and interleukin (IL)-6 expression in human HOBIT cells (Zheng et al. 1994), whereas BMP-7 increases IGF-I and -II expression in fetal rat calvarial cells. This induction of ALP by BMP-7 is reduced by IGF-I and -II antisense oligonucleotides,

suggesting that osteoblast differentiation induced by BMP-7 is mediated
by IGF expression (Yeh et al. 1996). BMP-7 also increases IGF binding
protein (IGFBP)-3 and -5 and decreases IGFB-4 expression in human
and rat osteoblastic cells (Gabbitas and Canalis 1995; Knutsen et al.
1995), suggesting that the actions of BMPs on osteoblastic cells are
mediated in part by IGFs and IGFBPs.

11.5 Mechanisms of Action of Bone Morphogenetic Proteins

Progress has been made recently in understanding the mechanisms of
action of BMPs on skeletal cells. Receptors binding BMP-2, -4 and -7
have been identified in osteoblastic cells (Malpe et al. 1994; Ten Dijke
et al. 1994). These BMP receptors belong to the family of serine/threon-
ine receptor kinases, which also include receptors for TGF-βs, activins
and inhibins (Massagué 1996). Two types of receptors are involved in
signaling by BMPs and TGF-β. After binding of the ligand to type II
receptors, type II BMP receptor associates with type I receptors to form
heterodimers. This causes phosphorylation of type I receptor and activation of the type I serine/threonine kinase activity (Massagué 1996).
Heterodimeric complexes consisting of type I and II receptors are
needed for signaling (Liu et al. 1995). Two types of type I BMP receptors have been identified, BMPR-IA/ALK3 and BMPR-IB/ALK6.
BMP-2, -4 and -7 bind to BMPR-IA and BMPR-IB with variable
affinities, depending on the presence of BMPR-II receptors (Ten Dijke
et al. 1994; Rosenzweig et al. 1995; Yamashita et al. 1996). BMP
receptors are expressed in distinct regions during embryonic skeletal
formation, suggesting specific functions of BMPs during embryogenesis (DeWulf et al. 1995). In the mouse embryo, both BMPR-IA and
BMPR-IB are expressed in condensing mesenchymal cells, in chondrocytes and osteoblasts. In the adult rat bone, only BMPR-IA is found in
periosteal osteoblasts. The expression of both BMPR-IA and BMPR-IB
is up-regulated in proliferating cells of the periosteum during fracture
healing, and both receptors are found in chondrocytes and osteoblasts in
the newly forming tissue, although BMPR-IA appears to be expressed at
a higher level than BMPR-IB in trabecular osteoblasts (Ishidou et al.
1995). The expression of BMPR-IA and BMPR-IB at sites of developing cartilage and bone supports a role for these receptors in the action of

BMPs in vivo (DeWulf et al. 1995). Interestingly, both BMPR-IA and -IB similarly induce ALP activity in the absence of BMP-2 in C2C12 myoblasts (Akiyama et al. 1997). The finding that BMP-2 can rapidly reduce TGF-β binding to type II receptors and increases binding to type I receptors in rat calvarial cells illustrates the complex interactions between BMP-2 and TGF-β receptors in osteoblastic cells (Centrella et al. 1995).

Recent data have brought new insights into the signaling cascades activated by the binding of BMPs to BMP receptors. In chicken limb micromass cultures, BMP-2 activates protein kinase A (PKA) activity and the level of cyclic AMP response element binding protein (CREB), leading to chondrogenesis, indicating that activation of PKA is involved in the induction of cartilage by BMP-2 (Lee and Chuong 1997). A mitogen-activated protein kinase kinase kinase (MAP-KKK), TAK1, was shown to transduce signals for BMP-4, but the downstream events are not yet known (Yamaguchi et al. 1995). Proteins related to the *Mad* gene family were recently found to be implicated in BMP-2 signal transduction (Hoodless et al. 1996). A human Mad protein (Smad1) encodes transcriptional activators induced by BMP-receptor-mediated signaling (Liu et al. 1996). Recent data indicate that Smad1 is phosphorylated directly and translocated to the nucleus after stimulation by BMP-2 (Hoodless et al. 1996; Kretzschmar et al. 1997). Both Smad1 and Smad5 appear to be involved in the intracellular BMP signals that inhibit myogenic differentiation and induce osteoblast differentiation in C2C12 cells (Yamamoto et al. 1997), further indicating that Mad family members act as transcription factors in the signal transduction pathway for BMPs.

The inductive effect of BMP-2 on osteoblast differentiation appears to be related to the induction of bHLH transcription factors (Tamura and Noda 1994). For example, BMP-2 regulates the expression of the *Id* gene in undifferentiated C3H10T1/2 cells and rat calvarial osteoblastic cells (Ogata et al. 1993). As discussed above, BMPs also modulate the expression of *Msx* genes. Recent data indicate that BMP-7 induces the expression of the osteoblast-specific transcription factor Osf2/Cbfa1 prior to the expression of other osteoblast marker genes, suggesting that this factor is part of the BMP signaling cascade in osteoblasts (Ducy et al. 1997). It is likely that BMPs act by modulating the expression of distinct transcription factors at different stages during mesenchymal cell differentiation.

11.6 Effects of Bone Morphogenetic Proteins on Bone Repair

As discussed above, BMPs were discovered because of their remarkable osteoinductive effect in vivo. Indeed, BMP-2, -3, -4,-5, -7, and more recently, BMP-6 and -9, were shown to induce the formation of bone after implantation in vivo (Reddi 1992, 1994; Wozney 1992; Riley et al. 1996; Rosen et al. 1996). The potential of BMPs to stimulate regeneration of bone has been demonstrated in various animal models (Riley et al 1996). Indeed, BMPs accelerate healing of segmental bone defects and non-unions in rats, dogs, rabbits and sheep, and promote bone healing in monkeys (Hollinger et al. 1989; Doll et al. 1990; Mark et al. 1990; Gerhart et al. 1993; Marden et al. 1993; Ripamonti et al. 1992; Muschler et al. 1994; Cook et al. 1994). In addition, BMP-2 was shown to induce or stimulate bone growth in femoral defects in rats and sheep, tibial and ulnar defects in rabbits, mandibular defects and spinal fusion in dogs and porous ingrowth in rats (Riley et al. 1996), whereas BMP-2, -3 and -4 promote periodontal regeneration (Giannobile 1996) and maxillofacial reconstruction (Boyne 1996). It is noteworthy that the choice of carrier material is crucial for the correct delivery of the proteins (Zellin et al. 1996). An important role of BMPs in the early stages of fracture repair is emphasized by the finding that BMP-2 and -4 are transiently expressed at fracture sites in mesenchymal cells, immature chondroblasts and pre-osteoblasts (Nakase et al. 1994; Bostrom et al. 1995).

11.7 Bone Morphogenetic Proteins and Osteoporosis

Besides determining the cellular and molecular mechanisms involved in the effects of BMPs on mesenchymal and skeletal cells, it is of obvious interest to determine whether BMPs may be of therapeutic value in humans. Given the potent induction of chondroblast/osteoblast differentiation in vitro and the local osteoinductive effects of BMPs in vivo, it can be postulated that BMPs may be useful for the treatment of clinical disorders characterized by insufficient bone formation. Bone formation declines with age in a variety of animal models, and BMPs may perhaps be used to stimulate bone formation and prevent bone loss occurring with aging. Although BMPs are able to induce ectopic bone formation in rats of all ages, the ectopic bone formation induced by BMPs diminishes with age (Irving et al. 1981; Jergesen et al. 1991; Fleet et al. 1996), presumably because of an age-related decline in BMP receptors in target cells. Although the activity of BMPs is compromised by advancing age, this can be partially reversed by increasing the dose at the implant site (Fleet et al. 1996). The age-related decrease in BMP responsiveness should be taken in account when considering using BMPs as a local therapeutic factor for bone regeneration in the elderly.

Few experimental studies have been conducted with BMPs to increase bone formation in experimental models of bone loss. In the young ovariectomized rat, a model of estrogen-deficient osteopenia, the reduced bone mass results from an increased bone resorption relative to bone formation (Modrowski et al. 1993). In this model, local infusion of TGF-β is not sufficient to increase bone formation (Beaudreuil et al. 1995). In contrast, we found that administration of IGF-I restores the reduced bone mass in old ovariectomized rats (Muller et al. 1994), reflecting the divergent actions of local anabolic factors on bone formation in vivo (Marie 1997a). Recent data indicate that the administration of BMPs in bone powder is ineffective in preventing bone loss in ovariectomized rats (Marquez et al. 1997). It could be of interest to determine whether BMPs may have beneficial effects in old ovariectomized rats where bone formation is markedly decreased.

In the model of bone loss induced by unloading in rats, we showed that the metaphyseal bone loss results from reduced osteoblastic cell proliferation leading to inhibition of bone formation (Machwate et al. 1993). In this model, we recently found that the systemic infusion of BMP-2 (2 µg/kg per day) had no effect on serum OC levels and histomorphometric indices of bone formation and did not prevent the decreased trabecular bone volume and bone mineral content induced by unloading, although BMP-2 stimulated the differentiation of rat marrow stromal osteoblastic cells in vitro (Zerath et al. 1998). In contrast, we found that systemic administration of IGF-I is effective in increasing bone formation and bone mass in this model (Machwate et al. 1994).

Morever, the systemic administration of TGF-β2 used at the same dose as BMP-2 increased osteoprogenitor cell proliferation, collagen type I transcripts and metaphyseal bone formation and prevented the trabecular bone loss induced by unloading in this model (Machwate et al. 1995). Thus, in contrast to the stimulation of cell proliferation induced by TGF-β2, stimulation of marrow stromal cell differentiation into osteoblasts by BMP-2 was not able to prevent the trabecular bone loss in this model of osteopenia. This indicates that promotion of osteoblastic cell proliferation and differentiation is better able to increase bone formation in vivo than is stimulation of osteoblast differentiation, in agreement with our previous studies in humans demonstrating that endosteal bone formation is mainly dependent on osteoblast precursor cell recruitment (Marie 1995).

11.8 Conclusions and Perspectives

Extensive basic research has shown that BMPs are involved in embryonic skeletal patterning, postnatal bone formation and repair. The coordinate actions of BMPs on mesenchymal cells during skeletal development and bone formation appear to be related to the differential expression of BMPs and their receptors in responsive cells and to interactions of BMPs with other regulatory factors. The cascade of biological events induced by BMPs leading to bone formation appears to result therefore from the coordinated action of different BMP-related proteins acting at different steps along the skeletal differentiation pathway. Although the cellular and molecular mechanisms of the actions of BMPs on skeletal cells are beginning to be understood, more research is needed to identify the signaling cascade and transcription factors involved in gene transcription induced by members of the BMP family. The development of such basic research in appropriate human models of osteoblast differentiation and osteogenesis may lead ultimately to the development of new drugs capable of increasing the actions of BMPs on skeletal cells in humans.

Some predictions can be made based on the available data. Concerning the clinical use of BMPs to improve bone repair, the development of new associations between biomaterials and BMPs may permit the optimal promotion of local bone formation in humans (Reddi 1994). However, , it will also be important to develop new methods for the local applications of BMPs. It could perhaps be possible to generate new bone in osteoporosis by promoting osteoprogenitor cell differentiation locally by increasing the expression of BMPs or BMP receptors in target cells. In vitro, the transfection of undifferentiated mesenchymal cells with genes encoding BMP-2 and BMP-4 induces the differentiation of osteogenic cells (Ahrens et al 1993). It should also be noted that cells expressing BMP-6 by genetic engineering are able to induce bone formation after implantation in mice (Gitelman et al. 1994). Recent data indicate that the implantation of degradable matrices containing plasmid DNA encoding BMP-4 leads to DNA uptake by fibroblasts and induction of rapid new bone formation and healing (Fang et al. 1996). This technology may provide an efficient way to deliver osteogenic factors in vivo at the correct location and at the right moment. We can then speculate that transfection of human stromal cells with appropriate vectors containing cDNA for BMPs or their receptors coupled to an osteoblast-specific promoter, followed by re-implantation of the cells at the site where new bone formation is required, may result in new bone formation. While this technology is available in vitro, several experimental and clinical studies have to be conducted to ensure that the correct BMPs and/or receptors are expressed at the desired time and place and that this leads ultimately to increased bone formation and bone mass. It is expected that the ongoing important developments in basic and clinical research into BMPs actions may provide new therapeutic applications to accelerate bone repair in humans, and, perhaps, to design safe therapeutic protocols aimed at promoting bone formation in patients with osteoporosis.

Acknowledgements. The author wishes to acknowledge past and present members of my group (F. Debiais, O. Fromigué, E. Hay, M. Hott, A. Lomri, M. Machwate, D. Modrowski) as well as collaborators (E. Zerath) who contributed to the studies reported in this review. Part of this work has been supported by INSERM, CNES and CNRS. BMP-2, IGF-I and TGF-β2 were kindly provided by Genetics Institute (USA) or Novartis (Basel, Switzerland).

References

Ahrens M, Ankenbauer T, Schroder D, Hollnagel A, Mayer H, Gross G (1993) Expression of human bone morphogenetic proteins-2 or –4 in murine mesenchymal progenitor C3H10T1/2 cells induces differentiation into distinct mesenchymal cell lineages. DNA Cell Biol 12:871–880

Akiyama S, Katagiri T, Namiki M, Yamaji N, Yamamoto N, Miyama K, Shibaya H, Ueno N, Wozney JM, Suda T (1997) Constitutively active BMP type I receptors transduce BMP-2 signals without the ligand in C2C12 myoblasts. Exp Cell Res 235:362–369

Amedee J, Bareille R, Rouais F, Cunningham N, Reddi H, Harmand, MF (1994) Osteogenin (bone morphogenic protein 3) inhibits proliferation and stimulates differentiation of osteoprogenitors in human bone marrow. Differentiation 58, 157–164

Asahina I, Sampath TK, Nisimura I, Hauschka PV (1993) Human osteogenic protein-1 induces both chondroblastic and osteoblastic differentiation of osteoprogenitor cells derived from newborn rat calvaria. J Cell Biol 123:921–933

Aubin JE, Turksen K, Heersche JNM (1993) Osteoblastic cell lineage. In: M Noda (ed) Cellular and molecular biology of bone. Academic, San Diego, pp 1–45

Beaudreuil J, Mbalavielé G, Cohen-Solal M, Morieux C, de Vernejoul MC, Orcel Ph (1995) Short term local injections of transforming growth factor-β1 decrease ovariectomy-stimulated osteoclastic resorption in vivo in rats. J. Bone Miner Res 10:971–977

Baylink DJ, Finkelman RD, Subburaman M (1993) Growth factors to stimulate bone formation. J Bone Miner Res 8:S565-S572

Benayahu D, Fried A, Shamay A, Cunningham N, Blumberg S, Wientroub S (1994) Differential effects of retinoic acid and growth factors on osteoblastic markers and CD10/NEP activity in stromal-derived osteoblasts. J Cell Biochem 56:62–73

Bitgood MJ, McMahon AP (1995) Hedgehog and BMP genes are cp-expressed at many diverse sites of cell-cell interaction in the mouse embryo. Dev Biol 172:126–138

Boden, SD, McCuaig K, Hair G, Racine M, Titus L, Wozney JM, Nanes MS (1996) Differential effects and glucocorticoid potentiation of bone morphogenetic protein action during rat osteoblast differentiation in vitro. Endocrinol 137:3401–3407

Bostrom MPG, Lane JM, Berberian WS, Missri AAE, Tomin E, Welland A, Doty SB, Glaser D, Rosen VM (1995) Immunolocalization and expression of bone morphogenetic proteins 2 and 4 in fracture healing. J Orthop Res 13:357–367

Boyne PJ (1996) Animal studies of application of rhBMP-2 in maxillofacial reconstruction. Bone 19:84S–91S

Canalis E and Gabbitas B (1994) Bone morphogenetic protein 2 increases insulin-like growth factor I and II transcripts and polypeptide levels in bone cell cultures. J Bone Miner Res 9:1999–2005

Carrington JL, Chen P, Yanaghishita M, Reddi AH (1991) Osteogenin (bone morphogenetic protein-3) stimulates cartilage formation by chick limb bud cells in vitro. Dev Biol 146:406–415

Centrella M, Casinghino S, Kim J, Pham T, Rosen V, Wozney J, McCarthy TL (1995) Independent changes in type I and type II receptors for transforming growth factor β induced by bone morphogenetic protein 2 parallel expression of the osteoblast phenotype. Mol Cell Biol 15:3273–3281

Chen D, Feng JQ, Harris MA, Mahy P, Mundy GR, Harris SE (1995) Sequence and expression of bone morphogenetic protein 3 mRNA in prolonged cultures of feral rat calvarial osteoblasts and in rat prostate adenocarcinoma PA III cells. DNA Cell Biol 14:235–239

Chen D, Harris MA, Rossini G, Dunstan SL, Dallas SL, Feng JQ, Mundy GR, Harris SE (1997) Bone morphogenetic protein 2 (BMP-2) enhances BMP-3, BMP-4, and bone cell differentiation marker gene expression during the induction of mineralized bone matrix formation in cultures of fetal rat calvarial osteoblasts. Calcif Tissue Inter 60:283–290

Chen P, Vukicevic S, Sampath TK, Luyten FP (1995) Osteogenic protein-1 promotes growth and maturation of chick sternal chondrocytes in serum-free cultures. J Cell Science 108: 105–114

Chen TL, Bates RL, Dudley A, Hammonds RG, Amento EP (1991) Bone morphogenetic protein-2b stimulation of growth and osteogenic phenotypes in rat osteoblast-like cells: comparison with TGF-β1. J Bone Miner Res 6:1387–1393

Cook SD, Baffes GC, Wolfe MW, SAmpath TK, Rueger DC (1994) Recombinant human bone morphogenetic protein-7 induces healing in a canine long-bone segmental defect model. Clin Orth Rel Res 301:302–312

Dallas SL, Park-Snyder S, Miyazono K, Twardzik D, Mundy GR, Bonewald LF (1994) Characterization and autoregulation of latent transforming growth factor beta (TGF beta) complexes in osteoblast-like cell lines. Production of a latent complex lacking the latent TGF-binding protein. J Biol Chem 269:6815–6821

Debiais F, Hott M, Graulet AM, Marie PJ (1998) Fibroblast growth factor-2 differently affects human neonatal calvaria osteoblastic cells depending on the stage of cell differentiation. J Bone Miner Res 13: 645–654

De Pollak C, Renier, Hott M, Marie PJ (1996) Increased bone formation and osteoblastic cell phenotype in premature cranial ossification (craniosynostosis). J Bone Miner Res 11:401–407

De Pollak C, Renier, Hott M, Marie PJ (1997) Temporal changes in bone formation, osteoblastic cell proliferation and differentiation during human postnatal calvarial suture ossification. J Cell Bioch 64:128–139

DeWulf N, Verschueren K, Lonnoy O, Moren A, Grimbsy S, Spiegle V, Miyazono K, Huylebroeck D, ten Dijke P (1995) Distinct spatial and temporal expression patterns of two type I receptors for bone morphogenetic proteins during mouse embryogenesis. Endocrinology 136:2652–2663

Dieudonné SC, Semeins CM, Goei SW, Vukicevic S, Nulend JK, Sampath TK, Helder M, Burger EH (1994) Opposite effects of osteogenic protein and transforming growth factor β on chondrogenesis in cultured long bone rudiments. J Bone Miner Res 9:771–780

Doll'BA, Towle HJ, Hollinger JO, Reddi AH, Mellonig JT (1990) The osteogenic potential of two composite graft systems using osteogenin. J. Periodontol. 61:745–750

Dollé P, Dierich A, LeMeur M, Schimmang T, Schuhbaur B, Chambon P, Duboule D (1993) Disruption of the Hoxd-13 gene induces localized heterochrony leading to mice with neotenic limbs. Cell 75:431–441

Ducy P, Zhang R, Geoffroy V, Ridall AL, Karsenty G (1997) Osf2/Cbf1: A transcriptional activator of osteoblast differentiation. Cell 89:747–754

Duprez D, Bell JH, Rivhardson MK, Archer CW, Wolpert L, Brickell PM, Francis-West PH (1996) Overexpression of BMP-2 and BMP-4 alters the size and shape of developing skeletal elements in the chick limb. Mech Dev 57:145–157

Fang J, Zhu YY, Smiley E, Bonadio J, Rouleau JP, Goldstein SA, McCauley LK, Davidson BL, Roessler BJ (1996) Stimulation of new bone formation by direct transfer of osteogenic plasmid genes, Proc Natl Acad Sci USA 93:5753–5758

Flechtenmacher J, Huch K, Thonar EJMA, Mollenhauer JA, Davies SR, Schmid TM, Puhl W, Sampath TK, Aydelotte MB, Kuettner KE (1996) Recombinant human osteogenic protein 1 is a potent stimulator of the synthesis of cartilage proteoglycans and collagens by human articular chondrocytes. Arthritis Rheum 39:1896–1904

Fleet JC, Casman K, Cox K, Rosen V (1996) The effects of aging on the bone inductive activity of recombinant human bone morphogenetic protein-2. Endocrinol 137:4605–4610

Francis PH, Richardson MK, Brickell PM, Tickle C (1994). Bone morphogenetic proteins and a signaling pathway that controls patterning in the developing chick limb. Development 120:209–218

Francis-West PH, Robertson KE, Ede DA, Rodriguez C, Izpisua-Belmonte JC, Houston B, Burt DW, Gribbin C, Brickell PM, Tickle C (1995) Expression of genes encoding bone morphogenetic proteins and sonic hedgehog in talpid (ta3) limb buds: their relationships in the signaling cascade involved in limb patterning. Dev Dyn 203:187–97

Fromigué O, Marie PJ, Lomri A (1997) Differential effects of transforming growth factor-β, 1,25-dihydroxyvitamin D and dexamethasone on human bone marrow stromal cells. Cytokine 9:613–623

Fromigué O, Marie PJ, Lomri A (1998) Bone morphogenetic protein-2 and transforming growth factor β2 interact to modulate human bone marrow stromal cell proliferation and differentiation. J Cell Biochem 68:411–426

Gabbitas B, Canalis E (1995) Bone morphogenetic protein-2 inhibits the synthesis of insulin-like growth factor-binding protein-5 in bone cell cultures, Endocrinology 136:2397–11

Gazit D, Ebner R, Kahn AJ, Derynck R (1993) Modulation of expression and cell surface binding of members of the transforming growth factor-β super-family during retinoic acid-induced osteoblastic differentiation of multipotential mesenchymal cells. Mol Endocrinol 7:189–198

Gerhart TN, Kirker-Head CA, Kriz MJ, Holtrop ME, Hennig GE, Hipp J, Schelling SH, Wang E (1993) Healing segmental femoral defects in sheep using recombinant human bone morphogenetic protein. Clin Orth Rel Res 293:317–326

Ghosh-Choudhury N, Harris MA, Feng JQ, Mundy GR, Harris SE (1994) Expression of the BMP2 gene during bone cell differentiation, Critical Rev Euk Gene Exp 4:345–355

Ghosh-Choudhury N, Windle JJ, Koop BA, Harris MA, Guerrero DL, Wozney JM, Mundy GR, Harris SE (1996) Immortalized murine osteoblast derived from BMP 2-T-antigen expressing transgenic mice. Endocrinology 137:331–339

Giannobile WV (1996) Periodontal tissue engineering by growth factors, Bone 19:23S-37S

Gitelman SE, Kobrin MS, Ye JQ, Lopez AR, Lee A, Derynck R (1994) Record Vgr-1/BMP-6-expressing tumors induce fibrosis and endochondral bone formation. J Cell Biol 126:1595–1609

Hanada K, Dennis JE, Caplan AI (1997) Stimulatory effects of basic fibroblast growth factor and bone morphogenetic protein-2 on osteogenic differentiation of rat marrow-derived mesenchymal stem cells. J Bone Miner Res 12:1606–1613

Harris SE, Harris MA, Feng JQ, Wozney J, Mundy GR (1994) Expression of bone morphogenetic protein messenger RNA in prolonged cultures of fetal rat calvaria cells. J Bone Miner Res 9:389–394

Harrison ET, Luyten FP, Reddi AH (1991) Osteogenin promotes reexpression of cartilage phenotype by differentiated articular chondrocytes in serum-free medium. Exp Cell Res 192:340–345

Hay E, Lomri A, Marie PJ (1998) Effects of rhBMP-2 on human neonatal calvaria osteoblastic cells J Cell Biochem (in press)

Hiraki Y, Inoue H, Shigeno C, Sanma Y, Bentz H, Rosen DM, Asada A, Suzuki
F (1991) Bone morphogenetic proteins (BMP-2 and BMP-3) promote
growth and expression of the differentiated phenotype of rabbit chondrocytes and osteoblastic MC3T3-E1 cells in vitro. J Bone Miner Res
6:1373–1385

Hogan BL (1996) Bone morphogenetic proteins in development. Curr Op Gen
Dev 6:432–438

Hofmann C, Luo G, Balling R, Karsenty G (1996) Analysis of limb patterning
in BMP-7 deficient mice. Dev Genetics 19:43–50

Hollinger J, Mark DE, Bach DE, Reddi AH, Seyfer AE (1989) Calvarial bone
regeneration using osteogenin. J Oral Maxillofac Surg, 47:1182–86

Hoodless PA, Haerry, T, Abdollah, S, Stapleton, M, O'Connor, MB, Atisano,
L, Wrana JL (1996) MADR1, a MAD-related protein that functions in
BMP2 signaling pathways. Cell 85:489–500

Hughes FJ, Collyer J, Stanfield M, Goodman SA (1995) The effects of bone
morphogenetic protein-2, -4, and -6 on differentiation of rat osteoblast cells
in vitro. Endocrinology 136:2671–2677

Inada, M, Katagiri T, Akiyawa S, Namika M, Komaki M, Yamaguchi A, Kamoi
K, Rosen V, Suda T (1996) Bone morphogenetic protein-12 and –13 inhibit
terminal differentiation of myoblasts, but do not induce their differentiation
into osteoblasts. Biochem Biophys Res Commun 222:317–22

Irving JT, LeBolt SA, Schneider EL (1981) Ectopic bone formation and aging.
Clin Orthop relat Res 154:249–253

Ishidou, Y, Kitajima I, Obama H, Maruyama I, Murata F, Imamura T, Yamada
N, Ten Dijke P, Miyazono K and Sakou T (1995) Enhanced expression of
type I receptors for bone morphogenetic proteins during bone formation. J
Bone Miner Res 10:1651–1657

Israel, DI, NoveJ, Kerns KM, Kaufman RJ, Rosen V, Cox KA, Wozney JM
(1996) Heterodimeric bone morphogenetic proteins show enhanced activity
in vitro and in vivo. Growth Factors 13: 291–300

Iwasaki, M, Nakahara H, Nakase T, Kimura T, Takaoka K, Caplan AI, Ono K
(1994) Bone morphogenetic protein 2 stimulates osteogenesis but does not
affect chondrogenesis in osteochondrogenic differentiation of periosteum-
derived cells. J Bone Miner Res 9:1195–1204

Jergesen HE, Chua J, Kao RT, Kalsau LB (1991) Age effects on bone induction by demineralized bone powder. Clin Ortop Rel Res 268:253–259

Katagiri, T, Yamaguchi A, Ikeda T, Yoshiki S, Wozney JM, Rosen V, Wang EA,
Tanaka H, Omura S, Suda T (1990) The non-osteogenic mouse pluripotent
cell line, C3H10T1/2, is induced to differentiate into osteoblastic cells by

recombinant human bone morphogenetic protein-2. Biochem Biophys Res Commun 172:295–299

Katagiri T, Yamaguchi A, Komaki M, Abe E, Takahashi N, Ikeda T, Rosen V, Wozney JM, Fujisawa-Schara, A, Suda T (1994) Bone morphogenetic protein-2 converts the differentiation pathway of C2C12 myoblasts into the osteoblast lineage. J Cell Biol 127:1755–1766

King JA, Marker PC, Seung KJ, Kingsley DM (1994) BMP5 and the molecular, skeletal, and soft-tissue alterations in short ear mice. Dev Biol 166:112–122

Kingsley DM (1994) The TGF-β superfamily: new members, new receptors, and new genetic tests of function in different organisms. Gene Dev 8:133–146

Knutsen, R, Mohan S, Wergedal J, Sampath K, Baylink DJ (1990) Osteogenic protein-1 stimulates proliferation and differentiation of human bone cells in vitro. J Bone Min Res 6: 231–238

Knutsen R, Honda Y, Strong DD, Sampath TK, Baylink DJ, Mohan S (1995) Regulation of insulin-like growth factor system components by osteogenic protein-1 in human bone cells. Endocrinology 136:857–865

Komaki M, Katagiri T, Suda T (1996) Bone morphogenetic protein-2 does not alter the differentiation pathway of committed progenitors of osteoblasts and chondroblasts. Cell Tiss Res 284:9–17

Kretzschmar M, Liu F, Hata A, Doody J, Massagué J (1997) The TGF-beta family mediator Smad1 is phosphorylated directly and activated functionally by the BMP receptor kinase. Genes Dev 11:984–995

Laufer E, Nelson CE, Johnson RL, Morgan BA, Tabin C (1994) Sonic hedgehog and FGF-4 act through a signaling cascade and feedback loop to integrate growth and patterning of the developing limb bud. Cell 79:993–1003

Lee YS, Chuong CM (1997) Activation of protein kinase A is a pivotal step involved in both BMP-2 and cyclic AMP-induced chondrogenesis. J Cell Physiol 170:153–165

Li IWS, Cheifetz S, McCulloch CAG, Sampath KT, Sodek J (1996) Effects of osteogenic protein-1 (OP-1, BMP-7) on bone matrix protein expression by fetal rat calvarial cells are differentiation stage specific. J Cell Physiol 169:115–125

Lind M, Eriksen EF, Bunger C (1996) Bone morphogenetic protein-2 but not bone morphogenetic protein-4 and -6 stimulates chemotactic migration of human osteoblasts, human marrow osteoblasts and U2-OS cells. Bone 18:53–57

Linkhart TA, Mohan, S, Baylink DJ (1996) Growth factors for bone growth and repair: IGF, TGF-β and BMP. Bone 19:1S–12S

Liu F, Ventura F, Doody J, Massagué J (1995) Human type II receptor for bone morphogenetic proteins (BMPs): extension of the two-kinase receptor model to the BMPs. Mol Cell Biol 15:3479–3486

Liu F, Hata A, B aker JC, Doody J, Carcamo J, Harland RM, Massagué J (1996) A human Mad protein acting as a BMP-regulated transcriptional activator. Nature 381:620–623

Lomri A, Marie PJ (1990) Bone cells responsiveness to TGF-β, parathyroid hormone and PGE_2 in normal and postmenopausal osteoporotic women. J Bone Miner Res. 5:1149–1155

Lomri A, de Pollak C, GoltzmAn D, Kremer R, Marie PJ (1997) Expression of PTHrP and PTH/PTHrP receptor in newborn human calvaria osteoblastic cells. Eur J Endocr 136:640–648

Luo G, Hofmann C, Bronckers ALJJ, Sohocki M, Bradley A, Karsenty G (1995) BMP-7 is an inducer of nephrogenesis, and is also required for eye development and skeletal patterning. Genes Dev 9: 2808–2820

Luyten FP, Cunningham NS, Ma S, Muthukumaran N, Hammonds RG, Nevins WB, Wood WI, Reddi AH (1989) Purification and partial amino acid sequence of osteogenin, a protein initiating bone differentiation. J Biol Chem 264:13377–13380

Luyten,FP, Yu YM, Yanagishita M, Vukicevic S, Hammonds RG, Reddi AH (1992) Natural bovine osteogenin and recombinant human bone morphogenetic protein-2B are equipotent in the maintenance of proteoglycans in bovine articular cartilage explant cultures. J Biol Chem 267:3691–3695

Lyons, KM, Pelton RW,Hogan LM (1989) Patterns of expression of murine Vgr-1 and BMP-2a RNA suggest that transforming growth factor-β-like genes coordinately regulate aspects of embryonic development, Genes Dev 3:1657–1668

Lyons, KM, Pelton RW, Hogan LM (1990) Organogenesis and pattern formation in the mouse: RNA distribution patterns suggest a role for bone morphogenetic protein-2 A (BMP-2 A). Development 109:833–844

Machwate M, Zerath E, Holy X, Hott M, Modrowski D, Malouvier A, Marie PJ (1993) Skeletal unloading inhibits bone formation and bone cell proliferation. Endocrinol Metab 264:E790–E799

Machwate M, Zerath E, Holy X, Hott M, Pastoureau P, Marie PJ (1994) Insulin-like growth factor-I increases trabecular bone formation and osteoblastic cell proliferation in unloaded rats. Endocrinolology 134, 3:1031–1038, 1994

Machwate M, Jullienne A, Moukhtar M, Lomri A, Marie PJ (1995) c-fos proto-oncogene is involved in the mitogenic effect of transforming growth factor-β in osteoblastic cells. Mol Endocrin 9:187–199

Machwate M, Zerath E, Holy E, Hott M, Godet D, Lomri A, Marie PJ (1995) Systemic administration of transforming growth factor-beta 2 prevents the impaired bone formation and osteopenia by unloading in rats. J Clin Invest 96:1245–1259

Macias D, Ganan Y, Sampath TK, Piedra ME, Ros MA, Hurle JM (1997) Role of BMP-2 and OP-1(BMP-7) in programmed cell death and skeletogenesis during chick limb development. Development 124:1109–1117

Maliakal, JC, Asahina I, Hauschka PV, Sampath TK (1994) Osteogenic protein-1 (BMP-7) inhibits cell proliferation and stimulates the expression of markers characteristic of osteoblast phenotype in rat osteosarcoma (17/2.8) cells. Growth Factors 11:227–234

Malpe, R, Baylink DJ, Sampath TK, Mohan S (1994) Evidence that human bone cells in culture contain binding sites for osteogenic protein-1, Biochem Biophys Res Commun 201:1140–1147

Marazzi G, Wang Y, Sassoon D (1997) Msx2 is a transcriptional regulator in the BMP-4-mediated programmed cell death pathway. Dev Biol 186:127–138

Marden LJ, Quighley NC, Reddi AH, Hollinger J (1993) Temporal changes during bone regeneration in the calvarium induced by osteogenin. Calcif. Tissue Int 53:262–268

Marie PJ, De Vernejoul MC (1993) Local factors influencing bone remodeling. Rev Rhum 601:55–63

Marie PJ, Hott M, Lomri A (1994) Regulation of endosteal bone formation and osteoblasts in rodent vertebrae. Cells and materials 4:143–154

Marie PJ (1995) Human endosteal osteoblastic cells: relationship with bone formation. Calcif Tissue Int 56:13–16

Marie PJ (1997a) Growth factors and bone formation in osteoporosis: roles for IGF-I and TGF-β. Rev Rhum 64:44–53

Marie PJ (1997b) Effects of bone morphogenetic proteins on cells of the osteoblastic lineage. J Cell Eng 2:92–99

Mark DE, Hollinger JO, Hastings C, Ma S, Chen G, Marden LJ, Reddi AH (1990) Repair of calvarial nonunions by osteogenin, a bone-inductive protein. Plast Reconst. Surg 86:623–630

Marquez JJ, Bald C, Martinez JA (1997) Bone metabolism and density in intact and ovariectomised rats after the administration of bone derived proteins. Cell Eng 2:107–112

Massagué J (1996) TGF-β signaling: receptors, transducers, and Mad proteins, Cell 85:947–950

Modrowski D, Miravet L, Feuga M, Marie PJ (1993) Increased proliferation of osteoblast precursor cells in estrogen deficient rats. Endocrinol Metab 27:E190-E196

Monsoro-Burq AH, Duprez D, Watanabe Y, Bontoux M, Vincent C, Brickell P, Le Douarin N (1996) The role of bone morphogenetic proteins in vertebral development. Development 122:3607–3616

Muller K, Cortesi R, Modrowski D, Marie PJ (1994) Stimulation of trabecular bone formation by insulin-like growth factor-I in adult ovariectomized rats. Am J Physiol 267: E1–E6

Mundy GR (1993) Cytokines and growth factors in the regulation of bone remodeling, J Bone Miner Res 8:S505–S510

Murray SS, Murray EJB, Glackin CA, Urist MR (1993) Bone morphogenetic protein inhibits differentiation and affects expression of helix-loop-helix regulatory molecules in myoblastic cells. J Cell Biochem 53:51–60

Muschler GF, Hyodo A, Manning T, Kambic H, Easley K (1994) Evaluation of human bone morphogenetic protein 2 in a canine spinal fusion model, Clin Orthop 308:229–240

Nakase T, Nomura S, Yoshikawa, H, Hashimoto J, Hirota S, Kitamura Y, Oikawa S, Ono K, Takaoka K (1994) Transient and localized expression of bone morphogenetic protein 4 messenger RNA during fracture healing. J Bone Miner Res 9:651–659

Nakashima, M, Nagasawa H, Yamda Y, Reddi AH (1994) Regulatory role of transforming growth factor-β, bone morphogenetic protein-2, and protein-4 on gene expression of extracellular matrix proteins and differentiation of dental pulp cells. Dev Biol 162:18–28

Nissinen L, Pirila L, Heino J (1997) Bone morphogenetic protein-2 is a regulator of cell adhesion. Exp Cell Res 230:377–385

Niswander L, Jeffrey S, Martin GR, Tickle C (1994) A positive feedback loop coordinates growth and patterning in the vertebrate limb. Nature 371:609–612

Ogata T, Wozney JM, Benezra R, Noda M (1993) Bone morphogenetic protein 2 transiently enhances expression of a gene, Id (inhibitor of differentiation), encoding a helix-loop-helix molecule in osteoblast-like cells. Proc Natl Acad Sci USA 90:9219–9222

Ogata, T, Wozney JM, Rodan GA, Noda M (1994) Bone morphogenetic protein-2 (BMP-2) acts both synergistically with and antagonistically against retinoic acid in regulating expression of phenotypic genes in osteoblast-like cells. Endocrine J 2:237–240

Owen ME (1985) Lineage of osteogenic cells and their relationships to the stromal cell system. Bone Miner 3:1–25

Ozkaynak, E, Schnegelsberg, PNJ, Jin DF, Clifford GM, Warren FD, Drier EA, Opperman H (1992) Osteogenic protein-2: a new member of the TGF-β superfamily expressed early in embryogenesis. J Biol Chem 267:25220–25227

Reddi, AH (1992) Regulation of cartilage and bone differentiation by bone morphogenetic proteins. Curr Op Cell Biol 4:850–855

Reddi AH (1994) Symbiosis of biotechnology and biomaterials: applications in tissue engineering of bone and cartilage. J Cell Biochem 56:192–195

Reddi AH (1995) Bone morphogenetic proteins, bone marrow stromal cells, and mesenchymal stem cells. Clin Orthop Rel Res 313:115–119

Reddi AH, Cunningham NS (1993) Initiation and promotion of bone differentiation by bone morphogenetic proteins. J Bone Miner Res 8:S499–S502

Rickard DJ, Sullivan TA, Shenker BJ, Leboy PS, Kazdhan I (1994) Induction of rapid osteoblast differentiation in rat bone marrow stromal cell cultures by dexamethasone and BMP-2. Dev Biol 161:218–228

Riley EH, Lane JM, Urist MR, Lyons KM, Lieberman JR (1996) Bone morphogenetic protein-2. Biology and applications. Clin Orthop Rel Res 324:39–46

Ripamonti U, Ma S, Reddi AH (1992) The critical role of geometry of porous hydroxyapatite delivery system in induction of bone by osteogenin, a bone morphogenetic protein. Matrix 12:202–212

Ripamonti U, Duneas N, Van den Heever B, Bosch C, Crooks J (1997) Recombinant transforming growth factor-β1 induces endochondral bone in the baboon and synergizes with recombinant osteogenic protein-1 (bone morphogenetic protein-7) to initiate rapid bone formation. J Bone Miner Res 12:1584–1595

Roark E, Greer K (1994) Transforming growth factor-β and bone morphogenetic protein-2 act by distinct mechanisms to promote chick limb cartilage differentiation in vitro. Dev Dyn 200:103–116

Rodan GA, Noda, M. (1991) Gene expression in osteoblastic cells. Crit Rev Euk Gene Exp 1:85–98

Rodriguez C, Kos R, Macias D, Abbott UK, Belmonte JCI (1996) Shh, HoxD, Bmp-2 and Fgf-4 gene expression during development of the polydactylous talpid2, diplopodia1, and diplopodia 4 mutant chick limb buds. Dev Gen 19:26–32

Rosen V, Thies RS (1992) BMPs in bone formation and repair. Trends Genet 8:97–102

Rosen V, Nove J, Song JJ, Thies RS, Cox K, Wozney JM (1994) Responsiveness of clonal limb bud cell lines to bone morphogenetic protein 2 reveals a sequential relationship between cartilage and bone cell phenotypes, J Bone Miner Res 9:1759–1768

Rosen V, Cox K, Hattersley G (1996) Bone morphogenetic proteins. In: JP Bilezikian, LG Raisz, GA Rodan (eds) Principles of bone biology, Academic, San Diego, 47:661–671

Rosenzweig BL, Imamura T, Okadome T, Cox GN, Yamashita H, Ten Dijke P, Heldin CH, Miyazono K (1995) Cloning and characterization of a human type II receptor for bone morphogenetic proteins. Proc. Natl. Acad. Sci. USA 92:7632–7636

Rudkin GH, Yamaguchi DT, Ishida K, Peterson WJ, Bahadosingh F, Thye D, Miller TA (1996) transforming growth factor-β, osteogenin, and bone morphogenetic protein-2 inhibit intercellular communication and alter cell proliferation in MC3T3-E1 cells. J Cell Physiol 168:433–441

Sampath, TK, Muthukumaran N, Reddi AH (1987) Isolation of osteogenin, an extracellular matrix-associated, bone-inductive protein, by heparin affinity chromatography, Proc Natl Acad Sci USA 84:7109–7113

Sampath, TK, Maliakal JC, Hauschka, PV, Jones WK, Sasak H, Tucker RF, White KH, Coughlin, JE, Tucker MM, Pang RHL, Corbett C, Ozkaynak E, Oppermann H, Rueger DC (1992) Recombinant human osteogenic protein-1 (hOP-1) induces new bone formation in vivo with a specific activity comparable with natural bovine osteogenic protein and stimulates osteoblast proliferation and differentiation in vitro, Proc Natl Acad Sci USA 267: 20352–20362

Storm, EE, Huynh, TV, Copeland NG, Jenkins NA, Kingsley, DM, Lee SJ (1994) Limb alterations in brachypodism mice due to mutations in a new member of the TGF-β superfamily. Nature 368:639–643

Takuwa, Y, Ohse C, Wang EA, Wozney JM, Uamashita K (1991) Bone morphogenetic protein-2 stimulates alkaline phosphatase activity and collagen synthesis in cultured osteoblastic cells, MC3T3-E1. Biochem Biophys Res Commun 174:96–101

Tamura M, Noda M (1994) Identification of a DNA sequence involved in osteoblast-specific gene expression via interaction with helix-loop-helix (HLH)-type transcription factors. J Cell Biol 126:3:773–782

Ten Dijke, P, Yamashita H, Sampath TK, Reddi AH, Estevez M, Riddle DL, Ichijo H, Heldin CH, Miyazono K (1994) Identification of type I receptors for osteogenic protein-1 and bone morphogenetic protein-4. J Biol Chem 269:16985–16988

Thies, RS, Bauduy M, Ashton BA, Kurtzberg L, Woznzy JM, Roen V (1992) Recombinant human bone morphogenetic protein-2 induces osteoblastic differentiation in W-20–17 stromal cells. Endocrinology 130:1318–1324

Thesleff I, Vaahtokari A, Partanen AM (1995) Regulation of organogenesis. Common molecular mechanisms regulating the development of teeth and other organs. Int J Dev Biol 39:35–50

Tickle, C (1994) On making a skeleton. Nature 368:587–588

Urist MR (1965) Bone formation by autoinduction. Science 150:893–899

Urist MR, Delange RJ, Finerman GAM (1983) Bone cell differentiation and growth factors: induced activity of chondro-osteogenic DNA. Science 220:680–686

Vainio S, Karanova I, Jowett A, Thesleff I (1993) Identification of BMP-4 as a signal mediating secondary induction between epithelial and mesenchymal tissues during early tooth development. Cell 75:45–58

Varghese S, Canalis E (1997) Regulation of collagenase-3 by bone morphogenetic protein-2 in bone cell cultures. Endocrinology 138:1035–1040

Vukicevic S, Luyten FP, Reddi AH (1989) Stimulation of the expression of osteogenic and chondrogenic phenotypes in vitro by osteogenin. Proc Natl Acad Sci USA 86:8793–8797

Vukicevic S, Paralkar VM, Cunningham NS, Gurkind JS, Reddi AH (1990a) Autoradiographic localization of osteogenin binding sites in cartilage and bone during rat embryonic development. Develop Biol 140:209–214

Vukicevic S, Luyten FP, Reddi AH (1990b) Osteogenin inhibits proliferation and stimulates differentiation in mouse osteoblast-like cells (MC3T3-E1). Biochem Biophys Res Commun 166:750–756

Wang EA, Rosen V, Cordes P, Hewick RM, Kritz MJ, Luxenberg DP, Sibley BS, Wozney JM (1988) Purification and characterization of other distinct bone-inducing factors. Proc Natl Acad Sci USA 85:9484–9488

Wang EA, Israel DI, Kelly S, Luxenberg DP (1993) Bone morphogenetic protein-2 causes commitment and differentiation in C3H10T1/2 and 3T3 cells. Growth Factors 9:57–71

Wang Y, Sassoon D (1995) Ectoderm-mesenchyme and mesenchyme-mesenchyme interactions regulate Msx-1 expression and cellular differentiation in the murine limb bud. Dev Biol 168: 374–382

Wozney JM (1992) The bone morphogenetic protein family and osteogenesis. Mol Rep Develop 32:160–167

Wozney JM, Rosen V, Celeste AJ, Mitsock LM, Whiters MJ, Kritz RW, Hewick RM, Wang EA (1988) Novel regulators of bone formation: molecular clones and activities. Science 242:1528–1534

Yamaguchi A (1995) Regulation of differentiation pathway of skeletal mesenchymal cells in cell lines by transforming growth factor-β superfamily. Cell Biol 6:165–173

Yamaguchi A, Ishizuya T, Kintou N, Wada Y, Katagiri T, Wozney JM, Rosen V, Yoshiki S (1996) Effects of BMP-2, BMP-4, and BMP-6 on onteoblastic differentiation of bone marrow-derived stromal cell lines, ST2 and MC3T3-G2/PA6. Biochem Biophys Res Commun 220:366–371

Yamaguchi A, Katagiri T, Ikeda T, Wozney JM, Rosen V, Wang EA, Kahn AJ, Suda T, Yoshiki S (1991) Recombinant human bone morphogenetic protein-2 stimulates osteoblastic maturation and inhibits myogenic differentiation in vitro. J Cell Biol 113:681–687

Yamamoto N, Akiyama S, Katagiri T, Namiki M, Kurokawa T, Suda T (1997) Smad1 and Smad5 act downstream of intracellular signalings of BMP-2 that inhibits myogenic differentiation and induces osteoblst differentiation in C2C12 myoblasts. Bioch Biophys Res Com 238:574–580

Yamashita H, Ten Dijke P, Heldin CH, Miyazono K (1996) Bone morphogenetic protein receptors. Bone 19:569–574

Yeh LCC, Adamo ML, Kitten AM, Olson MS, Lee C (1996) Osteogenic protein-1 mediated insulin-like growth factor gene expression in primary cultures or rat osteoblastic cells. Endocrinology 137:1921–1931

Zellin G, Hedner E, Linde A (1996) Bone regeneration by a combination of osteopromotive membranes with different BMP preparations: a review. Con Tissue Res 35:1–4; 279–284

Zerath E, Holy X, Noel B, Malouvier A, Hott M, Marie PJ (1998) Effects of BMP-2 on osteoblastic cells and on skeletal growth and bone formation in unloaded rats. Growth Hormone and IGF Res (in press)

Zheng MH, Wood DJ, Wysocki S, Papidimitriou JM, Wang EA (1994) Recombinant human bone morphogenetic protein-2 enhances expression of interleukin-6 and transforming growth factor-β1 genes in normal human osteoblast-like cells. J Cell Physiol 159:76–82

Zhou H, Hammonds BG, Findlay DM, Martin TJ, Ng KW (1993) Differential effects of transforming growth factor-β1 and bone morphogenetic protein 4 on gene expression and differentiated function of preosteoblasts. J Cell Physiol 155:112–119

Zou H, Niswander L (1996) requirement for BMP signaling in interdigital apoptosis and scale formation. Science 272:738–740

Subject Index

Ernst Schering Research Foundation Workshop

Editors: Günter Stock
Ursula-F. Habenicht